INTRODUCTION TO PSPICE®

A SUPPLEMENT TO

ELECTRIC CIRCUITS

FOURTH EDITION

JAMES W. NILSSON

PROFESSOR EMERITUS
IOWA STATE UNIVERSITY

SUSAN A. RIEDEL

MARQUETTE UNIVERSITY

INTRODUCTION TO PSPICE®

A SUPPLEMENT TO

ELECTRIC CIRCUITS

FOURTH EDITION

JAMES W. NILSSON

PROFESSOR EMERITUS
IOWA STATE UNIVERSITY

SUSAN A. RIEDEL

MARQUETTE UNIVERSITY

ADDISON-WESLEY PUBLISHING COMPANY

Reading, Massachusetts • Menlo Park, California
New York • Don Mills, Ontario • Wokingham, England
Amsterdam • Bonn • Sydney • Singapore
Tokyo • Madrid • San Juan • Milan • Paris

Senior Sponsoring Editor, Eileen Bernadette Moran
Production Supervisor, David Dwyer
Production Coordinator, Genevra Hanke
Copy Editor, Jerrold A. Moore
Text Designer, Sally Bindari, Designworks, Inc.
Illustrators, Capricorn Design
Cover Designer, Peter M. Blaiwas
Senior Manufacturing Manager, Roy Logan
Compositor, Interactive Composition Corporation

ISBN 0-201-51318-8

3 4 5 6 7 8 9 10-DO-959493

PREFACE

ABOUT THIS MANUAL

Introduction to PSpice® expressly supports the use of PSpice® (or SPICE) as part of an introductory course in electric circuit analysis based on the textbook *Electric Circuits, Fourth Edition*. This supplement focuses on three things: (1) learning to write PSpice programs; (2) constructing circuit models of basic devices such as op amps and transformers; and (3) learning to challenge computer output data as a means of reinforcing confidence in simulation. Because PSpice is designed to simulate networks containing integrated circuit devices, its range of application goes well beyond the topics covered in the textbook. Even though we do not exploit the full power of PSpice, we begin the introduction to this widely used simulation program at a level that you can use to test the computer solutions.

The use of PSpice involves learning a significant new technical vocabulary and a number of specialized techniques. Hence, we designed this supplement to stand on its own as an instructional unit. Our decision to separate this material from the parent textbook also is a service to you: The portable format greatly facilitates your use of the supplement at a computer terminal. In this format the supplement adds value, but not cost, to your study. This supplement is provided to purchasers of the text without any adjustment to the price of the textbook.

Introduction to PSpice contains 67 problems. We designed these problems with two objectives in mind: (1) to teach you how to construct PSpice source files; and (2) to teach you the importance of checking the output from PSpice simulation to ensure that the computer-generated solution makes sense in terms of known circuit behavior. Complete solutions to these problems are included in the solutions manual for *Electric Circuits*. You may also use PSpice to solve many of the textbook's Drill Exercises and chapter Problems.

INTEGRATING PSPICE INTO INTRODUCTORY CIRCUITS COURSES

Although some circuits courses cover PSpice as an independent topic, many instructors prefer to integrate computer solutions with the course. We used the following methods to support such integration: (1) notations in the margin of the textbook (marked in color by a logo and the acronym PSpice) refer to sections in the supplement that relate to the topic under study in the textbook; (2) topics appear in this supplement in the same order in which they are presented in the text; (3) notations in the margin of this manual (marked in color by a logo and the name Nilsson) refer to sections in the textbook that relate to the topic described in the manual; and (4) the accompanying table relates topics in the supplement to topics in the textbook.

ABOUT PSPICE

SPICE is a computer-aided simulation program that enables you to design a circuit and then simulate the design on a computer. SPICE is the acronym for a Simulation Program with Integrated Circuit Emphasis. The Electronics Research Laboratory of the University of California developed SPICE and made it available to the public in 1975.

Many different software packages are available that implement SPICE on personal computers or workstations. Among them, PSpice, from MicroSim Corporation, is the most popular. PSpice's popularity can be attributed to many factors, including its user-friendly interface, extensions to Spice that support modeling of digital circuits and much more, and its no-cost student version. This manual focuses on how to use PSpice, a superset of SPICE. We summarize differences between PSpice and SPICE in Appendix C and also mention them as appropriate in this supplement.

We limited our applications of PSpice to the types of circuit problems discussed in the textbook. Although PSpice is a general purpose program designed for a wide range of circuit simulation—including the simulation of nonlinear circuits, transmission lines, noise and distortion—here we discuss the use of PSpice only for dc analysis, transient analysis, steady-state sinusoidal (ac) analysis, and Fourier Series analysis.

MANUAL CHAPTER	TOPIC	PROBLEMS	TEXT CHAPTER
2–4	DC Analysis and Output	1–8	2
5	Op Amps	9–10	6
6–7	Transient Analysis	11–7	7–8
8	Switches	18–23	8
9	Step Response	24–32	8, 9
10	Varying Component Values	33–34	9
11	AC Steady-State Analysis	35–39	10
12	AC Power	40–42	11
13	Mutual Inductance	43–46	13
14	Ideal Transformers	47–48	13
15	Frequency Response	49–56	14, 17
16	Pulsed Sources	57–64	18
17	Fourier Series Analysis	65–67	18

CONTENTS

CHAPTER 1

PSPICE IN GENERAL

The general procedure for using PSpice consists of three basic steps. In the first step the user creates a source file for the circuit to be simulated or analyzed. In the second step the user enters the source file into the computer, which then runs the program and creates an output file. In the third and final step the user instructs the computer to print or plot the results of the output file.

Before we discuss creation of the source file, some general comments about the PSpice format are in order.†

1. Each statement in the source file may have several constituent parts. These parts, called fields, must be in a specific order but may appear anywhere in the line. Fields may be separated by one or more blanks, a comma, or a tab. In certain instances, equal signs or left and right parentheses are used as separators.

2. A statement in the source file cannot be longer than the 80 characters that fit in a single line on the computer display. A longer statement may be continued on the next line by including a plus sign (+) in column 1 of the next line, followed by the remainder of the statement.

3. PSpice does not distinguish between uppercase and lowercase characters. For example, "VIN", "Vin", and "vin" are equivalent. This usage does not hold for SPICE, where all characters in the source file must be uppercase—lowercase characters are not permitted. In this manual we use the convention of uppercase for PSpice keywords and lowercase for all other words. Thus if

† A. Vladimirescu, K. Zhang, A. R. Newton, D. O. Pederson, and A. Sangiovanni-Vincentelli, *SPICE Version 2G User's Guide,* University of California at Berkeley, unpublished; *PSpice Circuit Analysis, Version 5.0*. MicroSim Corporation, Irvine, Calif., 1991.

you attempt to use any of the example programs in this manual as inputs to SPICE, you must first convert characters to uppercase.

4. A name field must begin with a letter "a" through "z" (or "A" through "Z"), but the characters that follow may be letters, numbers, or any of the following: "$", "_", "*", "/", and "%". Names may be up to 131 characters long, but choosing a more reasonable length (say, 8 characters) is preferable. In SPICE, name fields may not contain any delimiters, and only the first 8 characters of the name are used.

5. A number field may be an integer (4, 12, −8) or a real number (2.5, 3.14159, −1.414). Integers and real numbers may be followed by either an integer exponent (7E–6, 2.136e3) or a symbolic scale factor (7U, 2.136k). Table 1 summarizes the symbolic scale factors used in PSpice and their corresponding exponential forms. Letters immediately following a number that are not scale factors are ignored, and so are letters immediately following a scale factor. For example, 10, 10v, 10HZ, and 10A all represent the same number, as do 2.5m, 2.5MA, 2.5msec, and 2.5MOhms.

6. The first statement in the source file is the title line, which may contain any type of text. The title line is ignored by PSpice, except to label the output upon completion of the analysis.

7. The last statement in the source file must be .END. Note that the period is part of the statement.

8. An "*" in the first column marks comment lines. Comments may contain any text and are ignored by PSpice.

9. Except for the title line, subcircuit definitions, the .OPTIONS statement with the .NOECHO parameter, and the .END statement, the order of the statements in the source file does not matter.

In creating a source file to direct PSpice to analyze a circuit, you must do three things. First, you must describe the circuit to be analyzed; second, you must state the type of analysis to be performed; and, third, you must specify the desired output. For convenience you should therefore divide the source file into three major subdivisions.

TABLE 1

PSPICE SCALE FACTORS

SYMBOL	EXPONENTIAL FORM	VALUE
F (or f)	1e–15	10^{-15}
P (or p)	1e–12	10^{-12}
N (or n)	1e–9	10^{-9}
U (or u)	1e–6	10^{-6}
M (or m)	1e–3	10^{-3}
K (or k)	1e3	10^{3}
MEG (or meg)	1e6	10^{6}
G (or g)	1e9	10^{9}
T (or t)	1e12	10^{12}

One subdivision consists of data statements that describe the circuit being simulated. Another comprises control statements that describe the type of analysis to be performed. The other contains the output specification statements, which control what outputs are to be printed or plotted. Although these subdivisions may be placed in any order in the source file, we will adopt the following convention in this manual: the data statements will appear first, followed by the control statements, followed by the output statements.

In addition to these major statement subdivisions, the source file must include two other statements. The *first* line of the source file must be a *title statement* and the *last* line of the source file must be an *end statement*.

You may create the source file with any text editor available on your system, so long as this editor does not insert any control characters. Most word processors make extensive use of control characters and so are generally not suited for creating PSpice source files. If your word processor has a configuration in which text can be created without the usual control characters, you may then create PSpice source files by using this configuration.

Also, if you are using PSpice within the Control Shell, the **Files** menu contains a text editor that you may use to create the source file. Working within the Control Shell has other significant advantages, such as the online help manual, which can assist in creating and modifying source files. See Appendix A for a detailed explanation of the Control Shell.

We now take a closer look at the three major subdivisions within a PSpice source file: the *data statements,* which describe the circuit; *control statements,* which specify the type of analysis and the *output statements,* which describe the type and format of the output.

1.1 THE DATA STATEMENTS

PSpice is based on nodal analysis. Therefore the first step in describing a circuit to PSpice is to number *all* the nodes in the circuit. The reference node *must* be numbered zero (0). The remaining nodes *must be numbered with nonnegative integers* but they need not be sequential. After numbering all the nodes in a circuit, you can describe a circuit completely by identifying the type of element connected to the nodes. In addition to describing the type of element, you must also specify its numerical characteristics. PSpice places several restrictions on the topological characteristics of a circuit. Therefore, when describing a circuit, you must make sure that (1) every node has at least two connections; (2) each node in the circuit has a dc path to the reference node; (3) the circuit does not contain loops of either voltage sources or inductors; and (4) the circuit does not contain cut sets of either current sources or capacitors.

The description of each passive circuit element and each independent source must contain three pieces of information: the type of element, two nodes to which it is connected, and its numerical value. The description of each controlled or dependent source must contain four pieces of information: the type of element, the two nodes to which it is connected, the controlling nodes, and its numerical value. The format for each element data statement consists of (1) the element name, (2) the circuit nodes to which the element is connected, and (3) the values of the parameters that describe the behavior of the element. The element name *must* start with a letter of the alphabet.

Nilsson Polarity references for voltage and current: Section 1.6, p. 15

Before concluding this overview of data statements, we need to pause and comment on polarity references. In PSpice, when polarity is relevant to the behavior of the element, the first node is positive with respect to the second node. This condition implies that the current reference direction is from the first-named node to the second-named node.

1.2 THE CONTROL STATEMENTS

As we mentioned earlier, PSpice was developed to analyze circuits containing integrated-circuit devices. One of the important aspects of such analysis is to determine the dc operating point, or bias level, of the circuit. As a result, PSpice automatically performs a dc analysis prior to both transient and ac analysis.

Therefore you may obtain a dc solution by just describing the circuit.

When other than *simple* dc analysis is desired, a control statement is used. The control statement consists of a PSpice command followed by fields that describe the parameters of the desired analysis. The number and type of these parameter fields depend on the type of analysis to be performed. The analysis commands described in this manual include `.DC` for complete dc analysis, `.AC` for ac analysis (frequency response), `.TRAN` for transient analysis (time response), `.FOUR` for Fourier Series analysis, `.OP` to compute the dc operating point of a circuit prior to other analysis, and `.TF` to compute the gain, input resistance, and output resistance (the transfer function). We present detailed descriptions of these commands later in this manual.

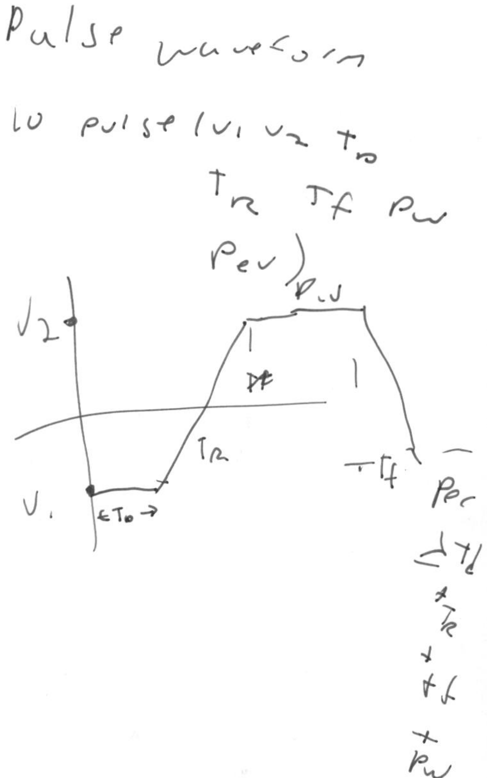

1.3 THE OUTPUT STATEMENTS

PSpice output has several constituent parts. The presence of each of these parts and the format used in the presentation is controlled by the output statements in the source file. We briefly describe these statements here and discuss them in more detail later in this manual.

The first section of the output file is the description of the circuit itself. At a minimum this section includes a listing of the source file created for PSpice to use in analysis.

The second section contains the default output from some of the analysis commands, which generate output without any specific directions from the source file. For example, the `.TF` command always causes the gain, the input resistance, and the output resistance to be printed in the output file after analysis is complete.

The third section contains the printout and line printer plots that have been explicitly requested in the source file. You use the `.PRINT` command to print data from the analysis. This statement includes fields that specify the type of analysis that generated the data and a list of the circuit variables whose values are to be printed. You use the `.PLOT` command to create line printer plots of data generated during PSpice analysis. The plot statement includes the same fields as the print statement.

The fourth section contains information that summarizes some statistical information describing the execution of PSpice. This information includes the memory requirements and the execution time and may be used in conjunction with the specification of some optional parameters to make your analysis more efficient.

In addition, the PSpice software is packaged with a graphics postprocessor called PROBE. It permits you to plot virtually any value of interest after PSpice has analyzed the circuit you have described. PROBE needs access to a special data file in order to function properly. You may generate this data file during PSpice analysis by including the .PROBE command in your source file. After PSpice has completed an error-free analysis of your circuit, you may activate PROBE either at the operating systems level or from the Command Shell. We describe PROBE in more detail later in this manual.

CHAPTER 2

DATA STATEMENTS FOR DC ANALYSIS

We begin our description of PSpice data statements with those that specify dc current and voltage sources. First, we present independent sources and then dependent sources. Whether independent or dependent, these sources are described in data statements consisting of four fields. The specific purpose of each field varies with the type of source being specified. We conclude with resistors, for which the data statements consist of three fields.

2.1 INDEPENDENT DC SOURCES

Data statements used for specifying independent dc sources include four fields. The first field is the element name, which must consist of a unique string of characters, the first letter of which identifies whether the element is a voltage source or a current source. If we let *xxx* denote an arbitrary alphanumeric string, then element names for independent sources must conform to the following convention:

Type of Element	**Element Name**
Independent voltage source	V*xxx*
Independent current source	I*xxx*

The second field contains the node numbers identifying the branch in which the dc source is located. The order in which the nodes are listed determines the polarity of the source. For a voltage source, the first node is positive with respect to the second node. For a current source, the current direction *through the source* is from the first node to the second node.

Nilsson Ideal independent voltage and current sources: Section 2.1, pp. 26–27

The third field identifies whether the source is constant or time varying, with a dc source indicated by DC in this field. Finally, the value of the dc source goes in the fourth field. For a voltage source, this value has the units volts, and for a current source the units are amperes.

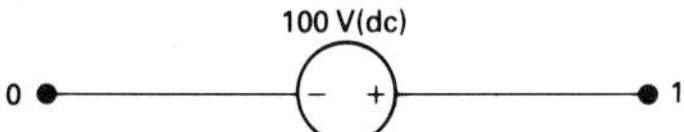

FIGURE 1 A 100-V dc source connected to nodes 1 and 0, with node 1 positive with respect to node 0.

For instance, the data statement for a 100-V dc voltage source named Vsourcel connected to nodes 1 and 0, as shown in Fig. 1, would be:

`Vsource1`	`1 0`	`DC`	`100`[†]
Name	Node connections	Type of source	Value

If we reverse the order of the nodes in the description of the source in Fig. 1, we enter the value as a negative number:

```
Vsource1  0 1  DC  -100.
```

In PSpice you may use an independent voltage source of zero value as an ammeter. That is, in order to measure a current, you may insert a zero-valued voltage source in series with the element where the current is to be determined. The zero-valued voltage source does not disturb the circuit because it is equivalent to a short circuit. These zero-valued voltage sources are necessary because the only current that PSpice is designed to output is the current through sources.

2.2 DEPENDENT DC SOURCES

Nilsson Ideal dependent voltage and current sources: Section 2.1, pp. 27–28

In discussing the data statements for dependent sources, we divide them into voltage-controlled sources and current-controlled sources. The description of a dependent source requires four fields. The first field contains the name of the source, with the first letter of the name identifying the type of dependent source. The second field contains the node numbers identifying the branch in which the source is located. The third field contains the node numbers of the controlling branch if the source is voltage controlled. If the source is current controlled the third field contains the name of the zero-valued voltage source used to compute the controlling current. The fourth field contains the *gain* of the dependent source, which is the ratio of the output

[†]This computer-like typeface is used here and in all other examples when the text can be typed verbatim in a PSpice source file.

quantity (voltage or current) to the controlling quantity (voltage or current).

For a voltage-controlled voltage source, the format is:

Exxx	N1 N2	NC1 NC2	VALUE[†]
Name	Node connections	Controlling nodes	Voltage gain

For instance, consider the circuit shown in Fig. 2, where the dependent voltage source is named ES2. The data statement for this voltage-controlled voltage source is:

ES2	4 3	1 2	3
Name	Node connections	Controlling nodes	Voltage gain

Note that the order of the nodes is relevant to the proper description of the source. Node 4 is positive with respect to node 3—hence the order 4, 3. The reference polarity for the controlling voltage v_Δ is positive at node 1; therefore the order of the controlling nodes is 1, 2.

The format of the data statement for a voltage-controlled current source is:

Gxxx	N1 N2	NC1 NC2	VALUE
Name	Node connections	Controlling nodes	Transconductance

The data statement for the voltage-controlled current source in the circuit shown in Fig. 3 and called GS04 is:

GS04	2 3	1 3	0.2
Name	Node connections	Controlling nodes	Transconductance

Here again the order of the nodes is essential to the proper description of the voltage-controlled current source. The reference direction for the current source is from node 2 to node 3; therefore the order of the node connections is 2, 3. The controlling voltage, v_Δ, is positive at node 1; thus the order of the controlling nodes is 1, 3.

A current-controlled source requires insertion of a zero-valued voltage source into the circuit to measure the controlling cur-

FIGURE 2 Circuit used to illustrate the data statement for a voltage-controlled voltage source.

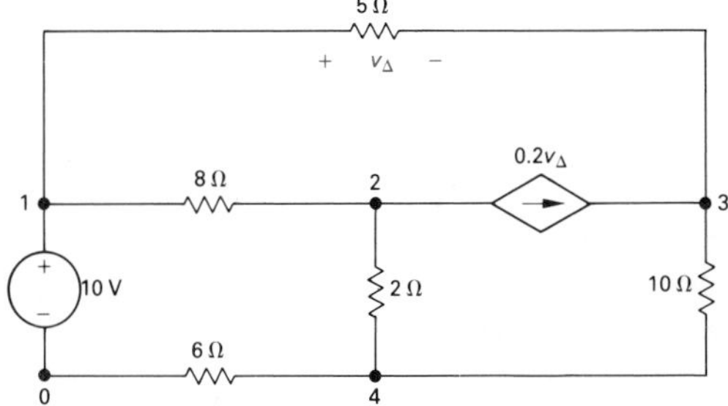

FIGURE 3 Circuit used to illustrate the data statement for a voltage-controlled current source.

[†] The normal uppercase typeface is used here and in all other examples when the source file elements are identified by descriptive placeholders. These placeholders cannot be typed verbatim in a PSpice source file. They must be replaced by values that PSpice expects.

rent. After you name this zero-valued voltage source, you enter it in the appropriate field in the data statement for the current-controlled source.

The format of the data statement for a current-controlled voltage source is:

H*xxx*	N1 N2	V*xxx*	VALUE
Name of current-controlled voltage source	Node connections	Name of the zero-valued voltage source used to measure the controlling current	Trans-resistance

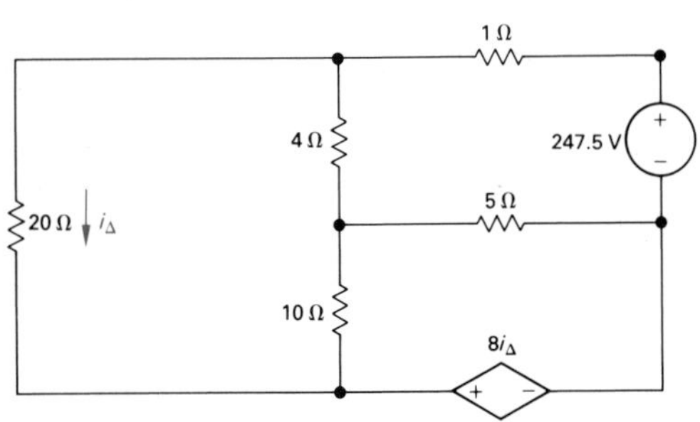

FIGURE 4 Circuit used to illustrate the data statement for a current-controlled voltage source.

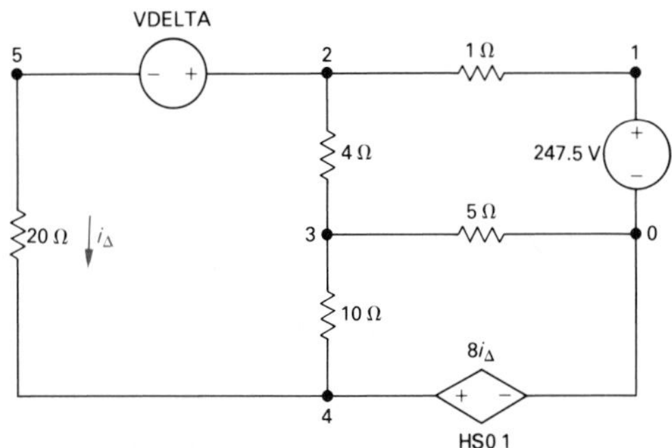

FIGURE 5 The circuit shown in Fig. 4, with the nodes numbered and the zero-valued voltage source inserted to measure the current i_Δ.

The polarity of the current-controlled voltage source is relevant to the behavior of the device; hence the order of the nodes is important. As before, N1 is positive with respect to N2. Note also that the controlling current is directed into the positive terminal of the zero-valued voltage source.

Consider the circuit shown in Fig. 4, which contains a current-controlled voltage source. Then consider the same circuit redrawn in Fig. 5, with the nodes numbered and a zero-valued voltage source labeled Vdelta inserted in series with the branch containing i_Δ. For the current-controlled voltage source named HS01, the pertinent data statement is:

`HS01`	`4 0`	`Vdelta`	`8`
Name	Node connections	Controlling current	Trans-resistance

The source file also contains the data statement describing the zero-valued voltage source:

`Vdelta`	`2 5`	DC	0
Name	Node connections	Type of source	Value

The format for the current-controlled current source is:

F*xxx*	N1 N2	V*xxx*	VALUE
Name of current-controlled current source	Node connections	Name of the zero-valued voltage source used to measure the controlling current	Current gain

Assume that the current-controlled current source in the circuit shown in Fig. 6 is labeled F1 and that the zero-valued voltage source used to measure the controlling current is called Valpha. Then the data statements describing the dependent source and the zero-valued independent source used to measure the current are:

```
F1       0 1   Valpha  0.1
Valpha   0 5   DC      0
```

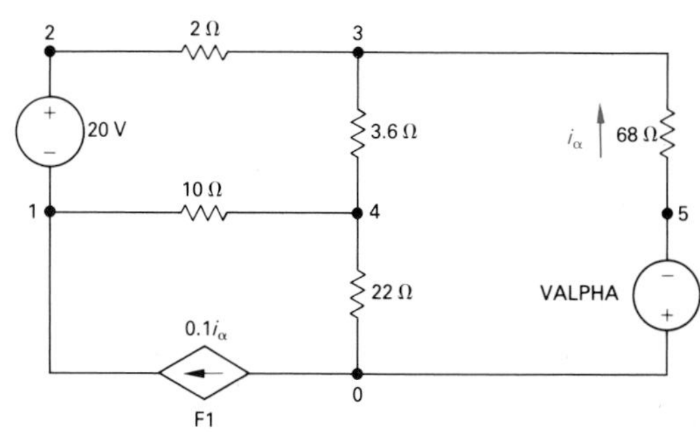

FIGURE 6 Circuit containing a current-controlled current source used to illustrate the data statements for this type of source.

2.3 RESISTORS

Data statements used for specifying resistors include three fields. The first is the element name, which must consist of a unique string of characters, the first of which is "R" (or "r") to identify the element as a resistor. The second contains the node numbers that identify the branch in which the resistor is located. The third contains the value of the resistance, in ohms.

Nilsson Resistors and Ohm's law: Section 2.2, pp. 29–32

The format for identifying a 100-Ω resistor named RX7 connected to nodes 7 and 12 is:

RX7	7 12	100
Name	Node connections	Value

We can now illustrate a PSpice source file for a resistive circuit by constructing one for the circuit shown in Fig. 5:

```
Example of a Resistive Circuit in PSpice
Ra        3  0   5
Rb        2  1   1
Rc        2  3   4
Rd        3  4   10
Re        4  5   20
Vdelta    2  5   DC      0
V1        1  0   DC      247.5
HS01      4  0   Vdelta     8
.END
```

We stated in Section 1.2 that you may perform a simple dc analysis without providing control statements or output statements in the source file. That is, only data statements that describe the circuit are needed. To illustrate this type of analysis we use PSpice to find the node voltages and source currents in the circuit shown in Fig. 6.

The input source file is:

```
Example of Simple DC Analysis
F1           0 1    Valpha     0.1
Valpha       0 5    DC         0
V1           2 1    DC         20
R1       1 4    10
R2       4 0    22
R3       2 3    2
R4       3 4    3.6
R5       3 5    68
.END
```

The pertinent printed output data are:

```
 NODE   VOLTAGE     NODE   VOLTAGE     NODE   VOLTAGE     NODE   VOLTAGE

(    1)  -14.1230  (    2)    5.8770  (    3)    3.2967  (    4)   -1.1732

(    5)    0.0000


    VOLTAGE SOURCE CURRENTS
    NAME          CURRENT

    Valpha       -4.848E-02
    V1           -1.290E+00

    TOTAL POWER DISSIPATION   2.58E+01  WATTS
```

Nilsson Definition of power for a basic circuit element: Section 1.6, pp. 14–15

Voltages are given in volts, and currents are given in amperes. Note that the PSpice output format uses "E" for the exponential. This is equivalent to "e" used in this manual.

The total power dissipation value is automatically printed whenever you invoke simple dc analysis. In PSpice this power represents the net power generated by the independent voltage sources in the circuit. If a dependent source or an independent current source is generating power, its value is not included. For example, in the circuit shown in Fig. 6 the power developed by the 20-V source is 25.80 W, and the total power dissipated in the five resistors is 25.87 W. The difference of 0.07 W is developed by the current-controlled current source. Example 1 presents a more dramatic illustration of this situation.

EXAMPLE 1

a) Use PSpice to find the voltages v_a and v_b for the circuit shown in Fig. 7.

b) Use the PSpice solutions to calculate (1) the total power dissipated in the circuit, (2) the power supplied by the independent current source, and (3) the power supplied by the current-controlled voltage source.

FIGURE 7 Circuit used to check power dissipation.

SOLUTION

a) The circuit shown in Fig. 7 is redrawn in Fig. 8. Note the insertion of a zero-value dc voltage source in series with the 20-Ω resistor. The names of the circuit components are shown to facilitate reading the PSpice source file:

```
Circuit Used to Check Power Dissipation
I1          0 1     DC          24
Vdelta      1 3     DC          0
H1     2 0      Vdelta      20.3846
R1     1 0      5
R2     3 0      20
R3     1 2      1
.END
```

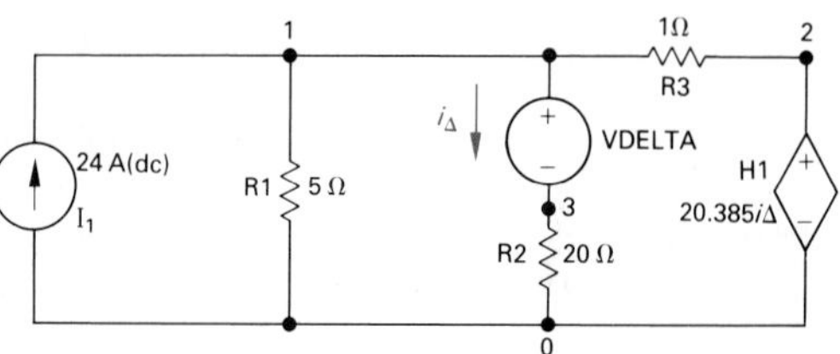

FIGURE 8 The circuit shown in Fig. 7 redrawn for PSpice analysis.

The pertinent PSpice output is:

```
NODE   VOLTAGE      NODE   VOLTAGE      NODE   VOLTAGE      NODE   VOLTAGE


(    1)  104.0000  (    2)  106.0000  (    3)  104.0000



     VOLTAGE SOURCE CURRENTS
     NAME              CURRENT

     Vdelta            5.200E+00

     TOTAL POWER DISSIPATION    0.00E+00   WATTS
```

Hence

$$v_a = V(1) = 104.00 \text{ V};$$

$$v_b = V(2) = 106.00 \text{ V}.$$

b) 1.

$$P_{5\Omega} = \frac{104^2}{5} = 2163.20 \text{ W};$$

$$P_{20\Omega} = \frac{104^2}{20} = 540.80 \text{ W};$$

$$P_{1\Omega} = \frac{(106 - 104)^2}{1} = 4.00 \text{ W}.$$

$$\Sigma P_{\text{dis}} = 2163.20 + 540.8 + 4 = 2708 \text{ W}.$$

2. P_{24A}(supplied) = 104(24) = 2496 W.

3. P_{H1}(supplied) = 106(2) = 212 W.

Note that the sum of power dissipated equals the sum of power supplied. Note also that the total power dissipated given by PSpice is 0 W, because there are no independent voltage sources in the circuit!

CHAPTER 3

CONTROL STATEMENTS FOR DC ANALYSIS

When other than the simple dc analysis described in Section 1.2 is desired, four dc control statements are available: .OP, .DC, .TF, and .SENS. (Note that control commands always begin with a period.)

3.1 THE .OP CONTROL STATEMENT

The .OP (operating point) control statement instructs PSpice to compute the bias point (the dc operating point) for the circuit. In fact, PSpice always calculates the operating point because this information is needed before you proceed with other types of analysis. Thus the actual effect of the .OP control statement is on the values printed in the output file. These values fall into three categories:

- voltages at each node;
- currents in each voltage source and total power dissipated; and
- operating points for each device.

The default output for simple dc analysis (described in Section 1.2) includes all values in the first two categories listed. Thus the .OP command causes operating point values to be printed for each device.

The values that define the operating point depend on the type of device. For example, the operating point values for dependent sources are the dc voltage across the device and the dc current through the device. If we add the .OP statement to the source file in Example 1, we can generate the following output in addition to the output produced by the source file used in Example 1:

```
**** CURRENT-CONTROLLED VOLTAGE SOURCES


NAME          H1
V-SOURCE       1.060E+02
I-SOURCE      -2.000E+00
```

From the circuit shown in Fig. 8 note that the voltage drop across the dependent source H1 is just the voltage at node 2, namely, Vsource = V(2) = 106 volts. The current through H1 is equal to the current flowing through R3 from node 1 to node 2 and can be computed from the node voltages as follows:

$$
\begin{aligned}
I_{\text{source}} &= \frac{V}{R3} \\
&= \frac{V(1) - V(2)}{1} \\
&= \frac{104 - 106}{1} \\
&= -2\,\text{A}.
\end{aligned}
$$

3.2 THE .DC CONTROL STATEMENT

The .DC (direct current) control statement allows you to increment an independent input source (either voltage or current) over a specified range of values. You also specify the size of the increment. The general format for the .DC control statement contains five fields:

.DC	SRCNAM	START	STOP	INCR
Command	Name of independent source	Starting value	Stopping value	Incrementing value

For example, the control statement

```
.DC  VS01  -5  10  1.0
```

causes a dc voltage source named VS01 to start at −5 V and stop at 10 V, with interim steps of 1.0 V. PSpice analyzes the circuit for each of these discrete values of VS01, so you might have to wait quite a long time before the analysis is completed.

The .DC control statement also allows you to sweep a second independent source over a specified range of values. For each specified value of the second source, you sweep the first source through its designated range. Hence, you can use PSpice to generate a family of characteristics; each member of the family corresponds to a specific value of the second source. Designers often use this feature of PSpice to plot the output characteristics of semiconductor devices.

When a second source enters the analysis, the format for the .DC control statement contains nine fields. The first contains .DC to identify the type of analysis. The next four describe the source name, starting value, ending value, and increment for the first source, just as these fields appear in the .DC statement for a single source. The last four describe the source name, starting value, ending value, and increment for the second source. The following statement represents symbolically the nine fields.

```
.DC   SRC1   START1   STOP1   INCR1   SRC2   START2
+   STOP2   INCR2
```

3.3 THE .TF CONTROL STATEMENT

The .TF (transfer function) control statement enables PSpice to compute three characteristics of the circuit being analyzed. First, it computes the ratio of the output variable to the input variable. This ratio is referred to as the *transfer function gain* of the circuit. Second, it computes the resistance with respect to the input source. Third, it computes the output resistance with respect to the terminals of the output element. The general format of the .TF control statement contains three fields:

.TF	OUTVAR	INSRC
Command	Output variable	Input source

The output file contains the same information obtained from a simple dc analysis and

1. the ratio OUTVAR/INSRC;
2. the input resistance with respect to INSRC; and
3. the output resistance with respect to OUTVAR.

Nilsson Determining Thévenin equivalents: Section 4.10, pp. 124–126

You may use the .TF control statement to find the Thévenin equivalent with respect to a designated pair of terminals. Examples 2 and 3 illustrate this application.

EXAMPLE 2

Use PSpice to find the Thévenin equivalent with respect to terminals a, b for the circuit shown in Fig. 9.

FIGURE 9 The circuit for Example 2.

SOLUTION

The circuit shown in Fig. 9 redrawn for PSpice analysis is shown in Fig. 10. The zero-value voltage source Vaib is used to measure the controlling current i_b. We labeled the components to facilitate reading the PSpice source file. Compare Fig. 10 with Fig. 9 and note that the Thévenin voltage v_{ab} is node voltage V(5, 0) or V(5). If we specify V(5, 0) as the output variable in the .TF control statement, the output resistance from PSpice is the Thévenin resistance.

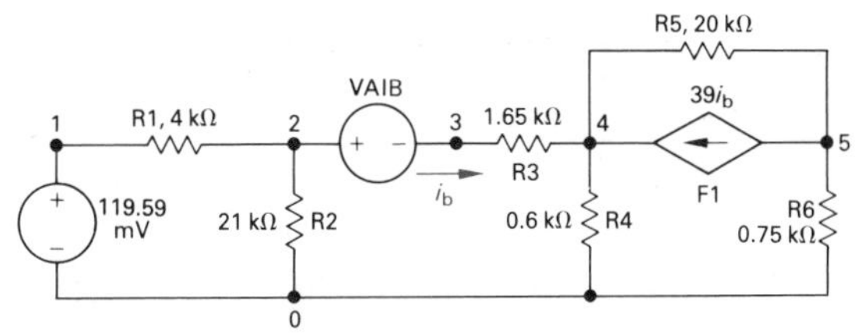

FIGURE 10 The circuit shown in Fig. 9 prepared for PSpice analysis.

The PSpice source file for the circuit in Fig. 10 is:

```
Finding a Thevenin Equivalent with PSpice
V1      1 0     DC      119.59e-3
R1      1 2     4e3
R2      2 0     21e3
Vaib    2 3     DC     0
R3      3 4     1.65e3
R4      4 0     600
R5      4 5     20e3
F1      5 4     Vaib      39
R6      5 0     750
.TF     V(5,0)      V1
.END
```

The pertinent data from the PSpice output is:

```
 NODE    VOLTAGE      NODE    VOLTAGE      NODE    VOLTAGE      NODE    VOLTAGE

 (    1)     .1196   (    2)     .0882   (    3)     .0882   (    4)     .0822

 (    5)    -.1000
```

```
****     SMALL-SIGNAL CHARACTERISTICS

      V(5,0)/V1 = -8.359E-01

      INPUT RESISTANCE AT V1 =   1.523E+04

      OUTPUT RESISTANCE AT V(5,0) =   7.446E+02
```

Hence the Thévenin equivalent is as shown in Fig. 11.

You may also calculate the Thévenin voltage from the transfer function ratio V(5)/V1:

$$\begin{aligned} V(5) &= \frac{V(5)}{V1} \cdot (V1) \\ &= (-0.8359)(0.11959) \\ &= -0.100. \end{aligned}$$

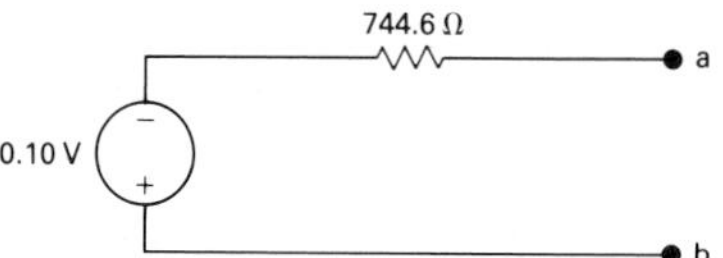

FIGURE 11 The Thévenin equivalent of the circuit shown in Fig. 10.

EXAMPLE 3

Use PSpice to find the Thévenin equivalent with respect to terminals a, b for the circuit shown in Fig. 12.

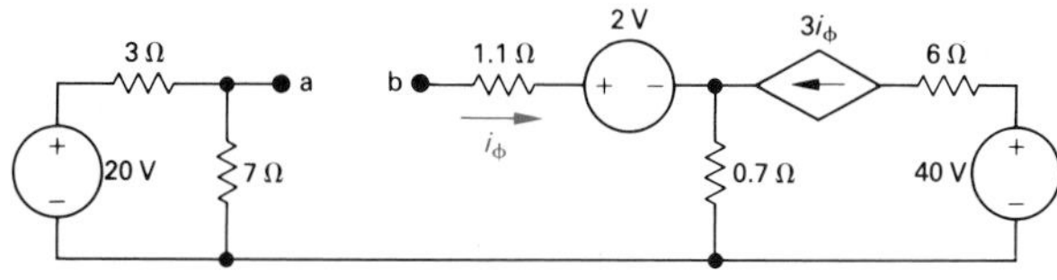

FIGURE 12 The circuit for Example 3.

SOLUTION

From the circuit shown in Fig. 12 note that i_ϕ is the current in an independent voltage source. Therefore we do not have to use a zero-value voltage source to measure the controlling current i_ϕ. Also note that there is only one connection to node b. PSpice requires at least two connections to every node, so we must modify the circuit accordingly. We may do so in one of two ways. First, we may connect a resistor that is large compared to the other resistors in the circuit—say, $10^6\,\Omega$ for this circuit—to node b and any other node in the circuit without affecting the behavior of the circuit. Second, we may connect a capacitor between node b and any other node. The capacitor behaves like an open circuit during dc analysis and therefore does not influence the dc Thévenin equivalent.

The circuit shown in Fig. 12 is redrawn in Fig. 13 in preparation for writing the PSpice source file. Note that we connected a resistor of $10^6\ \Omega$ between nodes 0 and 3. Also note that we chose node b as the reference node. We did so because PSpice automatically prints all the node-to-reference voltages, and therefore the Thévenin voltage becomes part of the printout if node b is the reference.

With three independent sources, we may choose which one to use in the .TF control statement. As we are interested only in the Thévenin voltage and resistance with respect to terminals a, b, the choice is immaterial. In other words, the transfer function ratio and the input resistance at the source are not relevant to the Thévenin equivalent at a, b. In the PSpice source file that follows we selected the 20-V source as the input source.

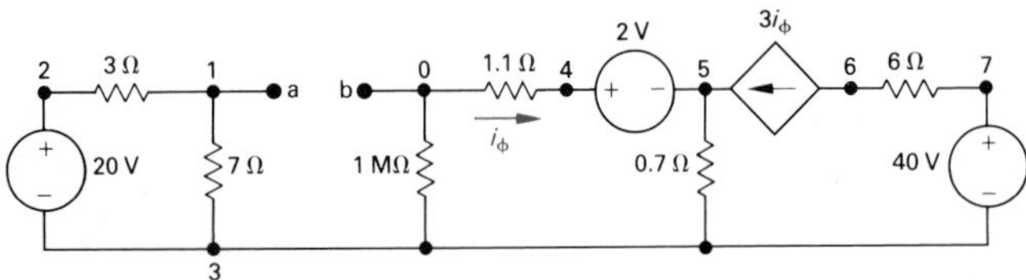

FIGURE 13 The circuit shown in Fig. 12 redrawn for PSpice analysis.

A PSpice source file for finding the Thévenin equivalent is:

```
Finding the Thevenin Equivalent when a Dependent Source is Present
R1      1 2      3
R2      1 3      7
V1      2 3      DC      20
R3      0 4      1.1
R4      0 3      1e6
V2      4 5      DC      2
R5      5 3      0.7
F1      6 5      V2      3
R6      6 7      6
V3      7 3      DC      40
.TF     V(1,0)          V1
.END
```

The relevant output from the analysis is:

```
 NODE    VOLTAGE      NODE    VOLTAGE      NODE    VOLTAGE      NODE    VOLTAGE

(    1)    12.0000  (    2)    18.0000  (    3)    -2.0000  (    4) 2.200E-06

(    5)    -2.0000  (    6)    38.0000  (    7)    38.0000
```

```
****      SMALL-SIGNAL CHARACTERISTICS

     V(1,0)/V1 =   7.000E-01

     INPUT RESISTANCE AT V1 =   1.000E+01

     OUTPUT RESISTANCE AT V(1,0) =   6.000E+00
```

Hence the Thévenin equivalent is a 14-V source in series with a 6-Ω resistor, as shown in Fig. 14.

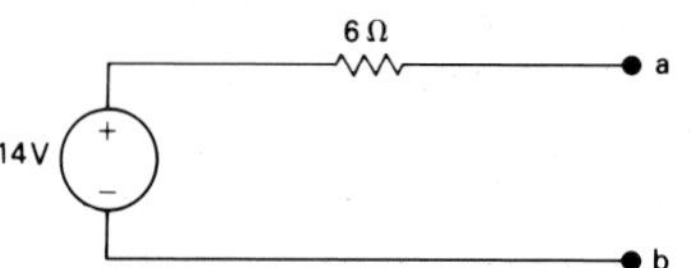

FIGURE 14 The Thévenin equivalent for Example 3.

3.4 THE .SENS CONTROL STATEMENT

The purpose of the . SENS (sensitivity) control statement is to obtain the dc small-signal sensitivities of each specified output variable with respect to *every* circuit parameter. Hence for circuits containing a large number of elements tremendous amounts of output data can be generated, particularly when the sensitivity of more than one output variable is of interest. Example 4 illustrates how to use the . SENS control statement to predict the behavior of an unloaded voltage-divider circuit.

EXAMPLE 4

Use PSpice to study the sensitivity of the output voltage v_o in the voltage-divider circuit shown in Fig. 15. We numbered the nodes in anticipation of writing a PSpice source file.

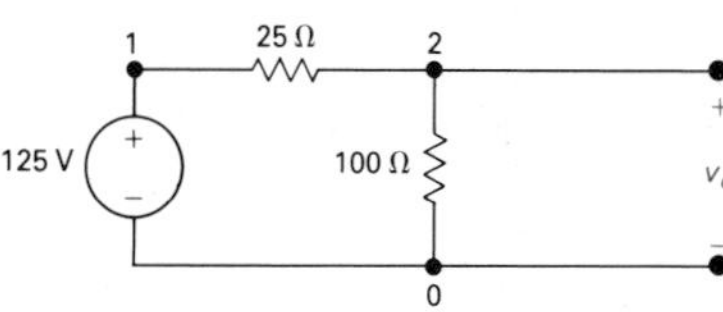

FIGURE 15 The circuit for Example 4.

SOLUTION

Because we have three elements in the circuit, PSpice calculates the sensitivity of v_o with respect to the independent voltage source, the 25-Ω resistor, and the 100-Ω resistor. The PSpice source file is:

```
Example of Sensitivity Analysis
V1     1  0     DC     125
R1     1  2     25
R2     2  0     100
.SENS     V(2,0)
.END
```

PSpice prints out the data associated with simple dc analysis and the sensitivity data. The data relevant to sensitivity are:

```
DC SENSITIVITIES OF OUTPUT V(2,0)

        ELEMENT          ELEMENT          ELEMENT           NORMALIZED
         NAME             VALUE         SENSITIVITY         SENSITIVITY
                                       (VOLTS/UNIT)      (VOLTS/PERCENT)

         R1             2.500E+01        -8.000E-01         -2.000E-01
         R2             1.000E+02         2.000E-01          2.000E-01
         V1             1.250E+02         8.000E-01          1.000E+00
```

From the simple dc analysis of the circuit, v_o = V(2) = 100 V when R1, R2, and V1 are at their nominal values. From the sensitivity data we deduce that

1. if R1 increases by 1 Ω, V(2) will decrease by 0.8 V—that is, V(2) = v_o = 99.2 V;
2. if R1 increases by 1%, v_o will decrease by 0.2 V to 99.8 V;
3. if R2 increases by 1 Ω, V(2) will increase by 0.2 V—that is, V(2) = v_o = 100.2 V;
4. if R2 increases by 1%, v_o will increase by 0.2 V to 100.2 V;
5. if V1 increases by 1 V, V(2) will increase by 0.8 V—that is, V(2) = v_o = 100.8 V; and
6. if V1 increases by 1%, v_o will increase by 1 V to 101 V.

Because we have a linear circuit, the principle of superposition applies, and therefore we may superimpose simultaneous effects. For example, let's assume that R1 increases by 1 Ω, R2 decreases by 1%, and V1 increases by 0.5 V. The effect on v_o, or V(2), will be $v_o = 100 - 0.8 - 0.2 + 0.4 = 99.4$ V.

CHAPTER 4

OUTPUT STATEMENTS FOR DC ANALYSIS

In Example 1 you could not specify which voltages and currents you want to examine in the output file. The .PRINT command gives you control over what appears in the output file.

The print statement has three fields. The first contains the command .PRINT. The second contains the type of analysis performed to generate the output—in this case, DC. The third contains a list of output variables to be printed. There is no limit to the number of output variables you can specify and no limit to the number of print statements in the source file.

```
.PRINT     DC         OV1  < OV2  OV3 . . . >        †
Command    Type of         Output
           analysis        variables
```

In order to use the print statement, you must understand the format for specifying voltage and current variables. Thus V(N1, N2) specifies the voltage difference between nodes N1 and N2. As before, the order of the nodes specifies the reference polarity. Omitting N2 causes the second node to default to zero or to the reference node.

Similarly, I(V*xxx*) specifies the current in an independent voltage source named V*xxx*. As before, positive current is *through the source*, from the positive node to the negative node.

Example 5 illustrates the use of the print statement in a source file containing a .DC control statement. Note that you must state explicitly the output variables to be printed by including a print statement (the .PRINT command) in the source file.

---

† Triangle brackets are used in examples of PSpice statements to identify optional fields. Ellipses are used in examples of PSpice statements to indicate additional identical fields. Neither triangle brackets nor ellipses should appear in actual PSpice statements.

EXAMPLE 5

For the circuit shown in Fig. 16 use PSpice to find the values of i_o and v_o when v_g varies from 0 to 100 V in 10-V steps.

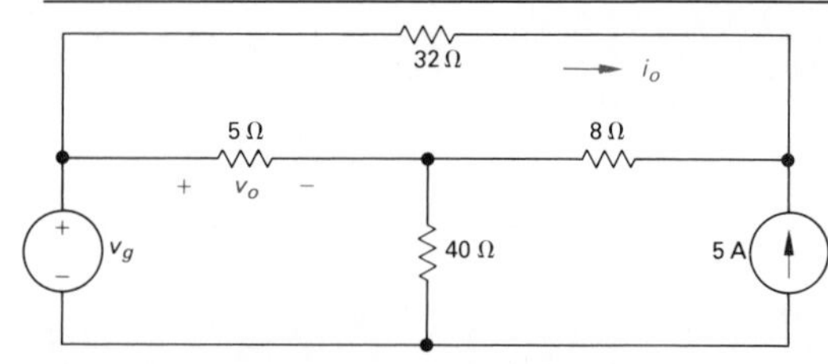

FIGURE 16 The circuit for Example 5.

SOLUTION

The circuit shown in Fig. 16 is redrawn in Fig. 17 with the nodes numbered and the elements named to facilitate reading the PSpice source file. Note the insertion of a zero-valued independent voltage source in series with the 32-Ω resistor to measure the current i_o. A PSpice source file that models this circuit is:

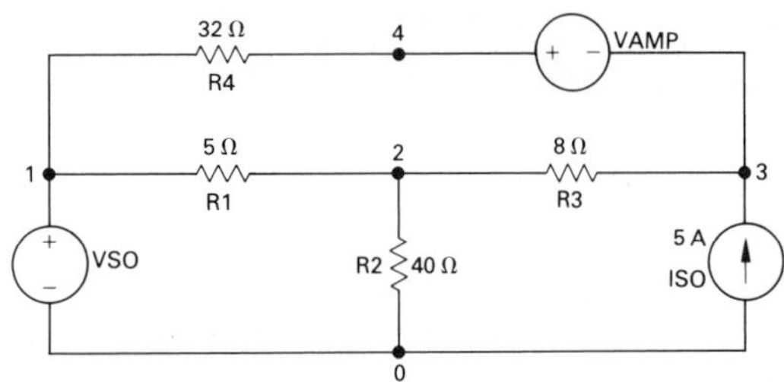

FIGURE 17 The circuit shown in Figure 16 prepared for PSpice analysis.

```
Circuit Used to Illustrate the .DC Control Statement
Vso     1 0     DC     0
Iso     0 3     DC     5
Vamp    4 3     DC     0
R1      1 2     5
R2      2 0     40
R3      2 3     8
R4      1 4     32
.DC     Vso     0      100     10
.PRINT     DC      I(Vamp)      V(1,2)
.END
```

The relevant output from the program is:

```
Vso            I(Vamp)        V(1,2)

 0.000E+00   -1.400E+00   -1.600E+01
 1.000E+01   -1.375E+00   -1.500E+01
 2.000E+01   -1.350E+00   -1.400E+01
 3.000E+01   -1.325E+00   -1.300E+01
 4.000E+01   -1.300E+00   -1.200E+01
 5.000E+01   -1.275E+00   -1.100E+01
 6.000E+01   -1.250E+00   -1.000E+01
 7.000E+01   -1.225E+00   -9.000E+00
 8.000E+01   -1.200E+00   -8.000E+00
 9.000E+01   -1.175E+00   -7.000E+00
 1.000E+02   -1.150E+00   -6.000E+00
```

Now we alter the source file of Example 5 to sweep through values of the current and voltage sources. Example 6 presents the results.

EXAMPLE 6

The current source in the circuit shown in Fig. 16 varies from 0 to 5 A in 1-A steps. For each value of the current tabulate v_o and i_g as v_g varies from 0 to 100 V in steps of 20 V.

SOLUTION

In order to print out Iso, we insert a zero-value voltage source in series with Iso. Refer to Fig. 17 and visualize this source placed between nodes 0 and 5, where node 5 is a new node inserted between Viso and Iso. The PSpice source file is:

```
Circuit with both Voltage and Current Sources Varied
Vso     1 0     DC      0
Iso     5 3     DC      0
Viso    0 5     DC      0
R1      1 2     5
R2      2 0     40
R3      2 3     8
R4      1 3     32
.DC     Vso     0       100     20      Iso     0       5       1
.PRINT     DC      V(1,2)      I(Viso)
.END
```

The output from the analysis is:

```
 Vso            V(1,2)         I(Viso)

  0.000E+00     0.000E+00      0.000E+00
  2.000E+01     2.000E+00     -1.840E-11
  4.000E+01     4.000E+00     -3.680E-11
  6.000E+01     6.000E+00     -5.520E-11
  8.000E+01     8.000E+00     -7.360E-11
  1.000E+02     1.000E+01     -9.200E-11
  0.000E+00    -3.200E+00      1.000E+00
  2.000E+01    -1.200E+00      1.000E+00
  4.000E+01     8.000E-01      1.000E+00
  6.000E+01     2.800E+00      1.000E+00
  8.000E+01     4.800E+00      1.000E+00
  1.000E+02     6.800E+00      1.000E+00
  0.000E+00    -6.400E+00      2.000E+00
  2.000E+01    -4.400E+00      2.000E+00
  4.000E+01    -2.400E+00      2.000E+00
  6.000E+01    -4.000E-01      2.000E+00
  8.000E+01     1.600E+00      2.000E+00
```

```
1.000E+02    3.600E+00   2.000E+00
0.000E+00   -9.600E+00   3.000E+00
2.000E+01   -7.600E+00   3.000E+00
4.000E+01   -5.600E+00   3.000E+00
6.000E+01   -3.600E+00   3.000E+00
8.000E+01   -1.600E+00   3.000E+00
1.000E+02    4.000E-01   3.000E+00
0.000E+00   -1.280E+01   4.000E+00
2.000E+01   -1.080E+01   4.000E+00
4.000E+01   -8.800E+00   4.000E+00
6.000E+01   -6.800E+00   4.000E+00
8.000E+01   -4.800E+00   4.000E+00
1.000E+02   -2.800E+00   4.000E+00
0.000E+00   -1.600E+01   5.000E+00
2.000E+01   -1.400E+01   5.000E+00
4.000E+01   -1.200E+01   5.000E+00
6.000E+01   -1.000E+01   5.000E+00
8.000E+01   -8.000E+00   5.000E+00
1.000E+02   -6.000E+00   5.000E+00


      JOB CONCLUDED

      TOTAL JOB TIME            2.14
```

Recall from Section 1.2 that in simple dc analysis the only output voltages printed are the node-to-reference voltages. You may use the .DC control statement and a print statement to obtain the voltage between any two nodes. You do so by sweeping only one source and then setting the sweep range to zero by making START1 = STOP1 and setting INCR to any positive value.

To illustrate we return to the circuit shown in Fig. 6. We already have a simple dc analysis of this circuit—see Example 1 in Section 2.3, with its accompanying printout. Recall that we obtained this printout without a .PRINT statement and that in this mode we have no control over what is printed. Let's assume that in the circuit shown in Fig. 6 the voltages V(3, 4), V(2, 4) and V(1, 4) are of interest. To obtain these voltages we modify that original PSpice source file in the following ways.

1. Change the independent voltage source entry to read

```
V1   2 1   DC   0
```

2. Add .DC control statement reading

```
.DC   V1   20   20   1
```

to the source file.

3. Add .PRINT statement requesting the desired voltages to the source file.

   The modified source file is now:

```
Example of Simple DC Analysis -- Revised
F1          0 1     Valpha     0.1
Valpha      0 5     DC         0
V1          2 1     DC         0
R1       1 4     10
R2       4 0     22
R3       2 3     2
R4       3 4     3.6
R5       3 5     68
.DC      V1      20      20      1
.PRINT      DC      V(3,4)      V(2,4)      V(1,4)
.END
```

The output voltages from the analysis are:

```
V1              V(3,4)          V(2,4)          V(1,4)

 2.000E+01    4.470E+00    7.050E+00   -1.295E+01
```

You should verify that these results are consistent with those given by the original source file. Note also that use of the .DC control statement results in printing only the requested node voltages, along with the voltage of the independent voltage source—V1 in this case.

CHAPTER 5

OPERATIONAL AMPLIFIERS

PSpice offers three options for describing an operational amplifier (op amp) within a source file. The first is to model the op amp with resistors and dependent sources. The second is similar to the first, but now the model is a PSpice subcircuit and is assigned a unique name. The op amp subcircuit may be used like any other PSpice circuit element, such as a resistor, and included in multiple locations within the circuit being described. The third is to take advantage of op amp models already supplied with PSpice. These models are available in a device library and may be accessed with the `.LIB` command. The library models are considerably more sophisticated than the simple dependent-source-based models and are designed to mimic the characteristics of actual op amps.

In general, if you have only one op amp in your circuit, modeling it with resistors and a dependent source is best. If the circuit contains two or more op amps, you should create an op amp subcircuit that you can use to represent each op amp in the circuit. You should use the library op amp models only when accurate modeling of certain op amp characteristics is important to the circuit analysis. Because the library models are complex, circuits containing them tend to occupy more space, and analysis of such circuits takes longer.

5.1 MODELING OP AMPS WITH RESISTORS AND DEPENDENT SOURCES

In Section 6.7 of the textbook the author introduced an equivalent circuit for the operational amplifier. That equivalent circuit

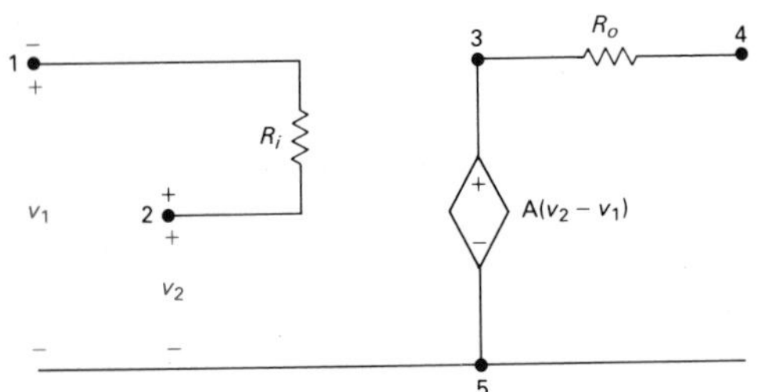

FIGURE 18 An equivalent circuit for an operational amplifier.

is redrawn in Fig. 18, where we also numbered the nodes to facilitate discussion of the PSpice model.

The PSpice description of the equivalent circuit shown in Fig. 18 is:

RI	1 2	VALUE OF R_i
E*xxx*	3 5 2 1	VALUE OF A
RO	3 4	VALUE OF R_o

Example 7 illustrates how to use PSpice to analyze a circuit containing an operational amplifier.

EXAMPLE 7

The parameters for the operational amplifier in the circuit shown in Fig. 19 are $R_i = 200$ kΩ, $A = 10^4$, and $R_o = 5$ kΩ.

a) Use PSpice to find v_i and v_o when $v_g = 1$ V(dc).

b) Compare the PSpice solution with the analytic solution for v_i and v_o.

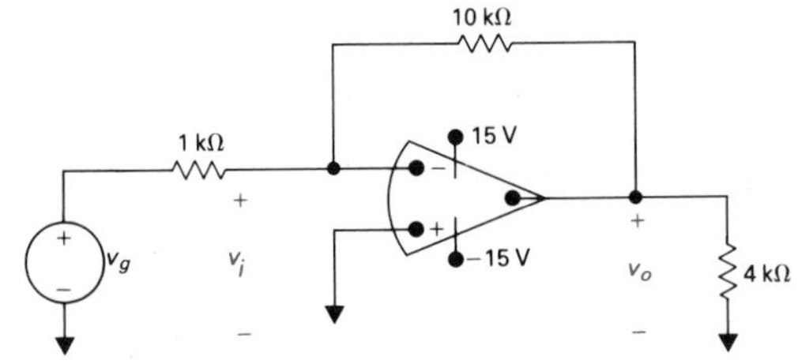

FIGURE 19 The circuit for Example 6.

SOLUTION

a) We redrew the circuit shown in Fig. 19 in Fig. 20 in preparation for writing the PSpice source file to find v_i and v_o:

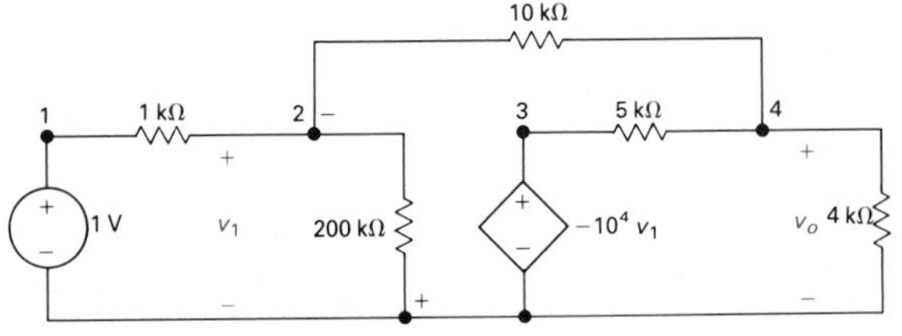

FIGURE 20 The circuit shown in Fig. 19 redrawn for PSpice analysis.

```
Example 6 -- The Op Amp
Vg        1 0       DC      1
Rs        1 2       1000
Ri        2 0       200e3
Rf        2 4       10e3
E1        3 0       2 0      -10e3
Ro        3 4       5e3
Rl        4 0       4e3
.END
```

The pertinent output from the PSpice program is

$$V(2) = 0.0027 \text{ V} = v_i;$$

$$V(4) = -9.969 \text{ V} = v_o.$$

b) We obtain the analytic solution for v_i and v_o by solving the following simultaneous node-voltage equations:

$$\frac{v_i - 1}{1} + \frac{v_i}{200} + \frac{v_i - v_o}{10} = 0; \quad \frac{v_o - v_i}{10} + \frac{v_o + 10^4 v_i}{5} + \frac{v_o}{4} = 0.$$

We leave to you verification that the solutions for v_i and v_o are

$$v_i = 2.742 \text{ mV} \qquad \text{and} \qquad v_o = -9.970 \text{ V}.$$

If we assume that the operational amplifier in the circuit shown in Fig. 19 is ideal, we can alter the PSpice op amp description by making R_i and A very large and $R_o = 0$. To illustrate we rewrite the source file in Example 7, with $R_i = 200$ MΩ, $A = 10^8$ and $R_o = 0$. By making $R_o = 0$, we eliminate node 3 from the circuit. The PSpice source file becomes:

```
Example 7 revisited -- The Ideal Op Amp
Vg          1 0      DC     1
Rs          1 2      1000
Ri          2 0      200e6
Rf          2 4      10e3
E1          4 0      2 0      -10e8
R1          4 0      4e3
.END
```

The values of v_i and v_o from this simulation are

$$v_i = \text{V}(2) = 0.0000 \text{ V};$$

$$v_o = \text{V}(4) = -10.0000 \text{ V}.$$

These results are consistent with the analysis of the circuit shown in Fig. 19, if we assume that the operational amplifier is ideal—that is, $v_i = 0$ V and $v_o = -10$ V.

5.2 MODELING OP AMPS WITH SUBCIRCUITS

In PSpice, a description of the topological connections of a circuit may include subcircuits. In other words, you may take a portion of an overall circuit, describe it separately, and then insert this description into the source file that describes the overall, or global, circuit. You define the subcircuit by naming the group of element statements that describe it. There is no limit on subcircuit size, and a subcircuit may contain other subcircuits.

You define a subcircuit in a PSpice source file by first inserting a . SUBCKT control statement, which has the general form:

```
.SUBCKT     SUBNAM     N1 N2 N3 . . .
```

where SUBNAM is the subcircuit name and N1 N2 N3 . . .

We test whether these results are reasonable by assuming that the op amps are ideal and then calculating the resulting values of v_{o1} and v_{o2}. We leave to you confirmation that these calculations yield $v_{o1} = -10$ V and $v_{o2} = -32$ V. We conclude from these results that we have made no errors in describing the circuit and hence are confident that the PSpice solution is correct.

5.3 USING OP AMP LIBRARY MODELS

One of the most powerful features of PSpice is the library of models for various electronic devices. The full version of PSpice has models for more than 5000 devices! The student version of PSpice has models for about 30 of the most commonly used devices. Among these models are two for op amps: the LM324 and the μA741. Figure 23 presents a listing of the subcircuit model for the μA741. As you can see, this model is very complex, and some of its components are well beyond the scope of this manual or the textbook. Note that the initial comments describe the external nodes for the subcircuit.

Nilsson Description of the μA741 op amp: Section 6.1, pp. 190–191

In order to incorporate the μA741 subcircuit into your circuit you must tell PSpice which library contains the model you want to use. As illustrated in Example 8, you use the .LIB command in this case, with the format

```
.LIB        fname.lib
```

where fname is the name of the library file.

```
*-----------------------------------------------------------------------------
* connections:    non-inverting input
*                 | inverting input
*                 | | positive power supply
*                 | | | negative power supply
*                 | | | | output
*                 | | | | |
.subckt uA741     1 2 3 4 5
*
  c1    11 12 8.661E-12
  c2     6  7 30.00E-12
  dc     5 53 dx
  de    54  5 dx
  dlp   90 91 dx
  dln   92 90 dx
  dp     4  3 dx
  egnd 99   0 poly(2) (3,0) (4,0) 0 .5 .5
```

FIGURE 23 The μA741 Op Amp Subcircuit Model from the PSpice library file.

```
  fb      7 99 poly(5) vb vc ve vlp vln 0 10.61E6 -10E6 10E6 10E6 -10E6
  ga      6  0 11 12 188.5E-6
  gcm     0  6 10 99 5.961E-9
  iee    10  4 dc 15.16E-6
  hlim   90  0 vlim 1K
  q1     11  2 13 qx
  q2     12  1 14 qx
  r2      6  9 100.0E3
  rc1     3 11 5.305E3
  rc2     3 12 5.305E3
  re1    13 10 1.836E3
  re2    14 10 1.836E3
  ree    10 99 13.19E6
  ro1     8  5 50
  ro2     7 99 100
  rp      3  4 18.16E3
  vb      9  0 dc 0
  vc      3 53 dc 1
  ve     54  4 dc 1
  vlim    7  8 dc 0
  vlp    91  0 dc 40
  vln     0 92 dc 40
.model dx D(Is=800.0E-18 Rs=1)
.model qx NPN(Is=800.0E-18 Bf=93.75)
.ends
```

EXAMPLE 8

Repeat the analysis of the circuit shown in Fig. 22, but this time use the μA741 model from the PSpice model library.

SOLUTION

The model's source file listing shows that the μA741 subcircuit model requires explicit connections to the positive and negative voltage supplies. These supplies are called VCC1 and VEE1 for the left-hand op amp, and VCC2 and VEE2 for the right-hand op amp.

```
Using Library Op Amp Models
VG          1 0        DC      0
RS1         1 2        4E3
RF1         2 3        80E3
R1          3 4        5E3
R2          4 0        20E3
RS2         5 0        6E3
RF2         5 6        18E3
RL          6 0        10E3
X1          0 2 7 8 3      uA741
X2          4 5 9 10 6     uA741
VCC1        7 0        DC      10
VEE1        0 8        DC      10
```

```
VCC2        9 0      DC     40
VEE2        0 10     DC     40
.LIB    EVAL.LIB
.DC     VG      0.5      0.5     1
.PRINT      DC      V(3)      V(6)
.END
```

The output from this source file is

$$v_{o1} = \mathrm{V}(3) = -9.577 \text{ V};$$

$$v_{o2} = \mathrm{V}(6) = -30.64 \text{ V};$$

Note that these answers are slightly different from those obtained when the simplified subcircuit model was used in Section 5.2.

CHAPTER 6

INDUCTORS AND CAPACITORS

Data statements for inductors and capacitors are similar to those used for resistors. As a minimum, three fields are required. The first consists of a unique element name. If the element is an inductor, the name must begin with the character "L" (either uppercase or lowercase); if the element is a capacitor, the name must begin with the character "C" (either uppercase or lowercase). As with the description of a resistor, the second field contains the node numbers that describe the branch in which the inductor or capacitor is located. The third field contains the value of the inductance (in henries) or the capacitance (in farads).

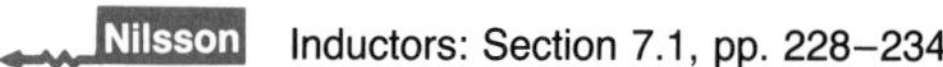
Nilsson Inductors: Section 7.1, pp. 228–234

A 10-mH inductor named LPGA8 connected between nodes 3 and 9 is specified as:

```
LPGA8     3 9          10e-3
 Name     Node         Value
          connections
```

The format for a 0.58-μF capacitor denoted C1 and connected between nodes 5 and 0 is specified as:

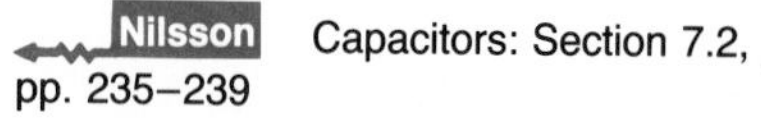
Nilsson Capacitors: Section 7.2, pp. 235–239

```
C1        5 0          0.58e-6
Name      Node         Value
          connections
```

You do not have to give the reference node 0 explicitly. That is, if you give only one node number, the second node defaults to 0.

For cases where inductors carry initial current, the current reference direction is from the first-named node to the second-named node. For instance, consider the inductor shown in Fig. 24, which is carrying an initial current of 5 A oriented from

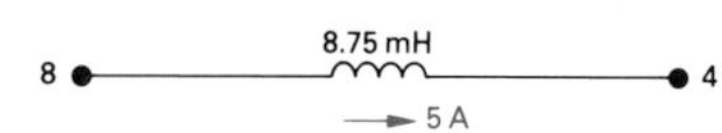

FIGURE 24 An inductor carrying initial current.

node 8 to node 4. Let's assume that the inductor is named L5. The element data statement in the source file then is:

L5	8 4	8.75e-3	IC = 5
Name	Node connections	Value	Initial condition

Reversing the order of the nodes in the description of L5 requires that the initial current be given as a negative value. In other words, we may also describe the inductor in Fig. 24 as:

```
L5   4 8   8.75e-3   IC = -5.
```

For cases where capacitors carry initial voltage, the polarity of that voltage is positive at the first-named node. For instance, consider the capacitor shown in Fig. 25, which is carrying an initial voltage of 12.8 V, positive at node 12. Assume that the capacitor is named Cap36. The element data statement in the source file then is:

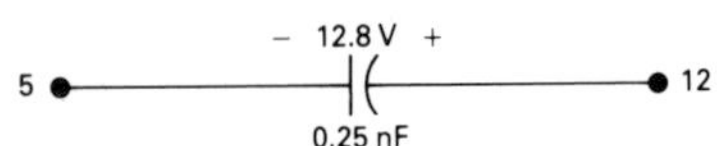

FIGURE 25 A capacitor carrying an initial voltage.

Cap36	12 5	0.25e-9	IC = 12.8
Name	Node connections	Value	Initial condition

As in the preceding case, reversing the order of the nodes requires entering the initial condition as a negative value. Hence we may also describe the capacitor in Fig. 25 as

```
Cap36   5 12   0.25e-9   IC = -12.8.
```

CHAPTER 7

TIME-DEPENDENT CIRCUIT RESPONSE

We now describe some of the features of PSpice that you can use when examining the behavior of a circuit as a function of time. In PSpice, this behavior is called the transient response of the circuit.

7.1 EXPONENTIAL PULSE SOURCES

In order to facilitate transient analysis, PSpice provides independent voltage and current sources that are time dependent. In this section we describe the exponential pulse. Although we base this description on an independent voltage source, we may also use this format to describe an independent current source.

The exponential voltage pulse is described by the following equations:

$$v_g = \mathrm{V1}, \qquad 0 \le t \le \mathrm{TD1};$$

$$v_g = \mathrm{V1} + (\mathrm{V2} - \mathrm{V1})(1 - e^{-(t-\mathrm{TD1})/\tau_1}), \quad \mathrm{TD1} \le t \le \mathrm{TD2};$$

$$v_g = \mathrm{V1} + (\mathrm{V2} - \mathrm{V1})(1 - e^{-(t-\mathrm{TD1})/\tau_1}) + (\mathrm{V1} - \mathrm{V2})(1 - e^{-(t-\mathrm{TD2})/\tau_2}) \qquad \mathrm{TD2} \le t \le \mathrm{TSTOP};$$

where

V1 = initial value in volts;
V2 = pulsed value in volts;
TD1 = rise delay time in seconds;
τ_1 = rise-time constant in seconds;
TD2 = fall delay time in seconds; and
τ_2 = fall-time constant in seconds.

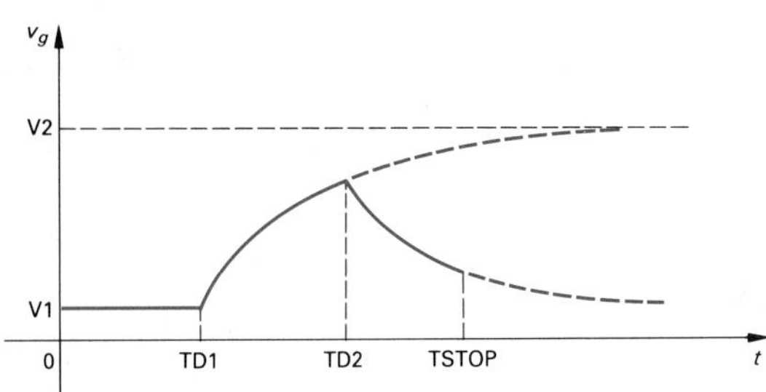

FIGURE 26 The exponential voltage pulse.

The default values are as follows: TD1 = 0, τ_1 = TSTEP,TD2 = TD1 + TSTEP, and τ_2 = TSTEP. Figure 26 shows a graph of the exponential voltage pulse.

The syntax for an exponential voltage source, called Vexpso, positive at node N1 and connected to nodes N1 and N2, is

Vexpso N1 N2 EXP(V1 V2 TD1 τ_1 TD2 τ_2).

7.2 TRANSIENT ANALYSIS

You use transient analysis to examine the response of a circuit as a function of time. The control statement for transient analysis has six fields, the last three of which are optional. The format for the transient analysis statement is:

.TRAN TSTEP TSTOP < TSTART < TMAX >>

\+ < UIC >

The terms inside the angle brackets <> are optional, as we discuss shortly. TSTEP is the printing or plotting increment. TSTOP is the final time. TSTART is the starting time; if omitted, it is assumed to be zero. TMAX is the maximum step size that PSpice uses for computational purposes. If TMAX is omitted, the value used defaults to either TSTEP or (TSTOP − TSTART)/50, whichever is smaller. TMAX is useful when you want to guarantee a computing interval smaller than the printing or plotting interval (i.e., TSTEP).

Transient analysis always begins at time zero. If TSTART is not zero, PSpice analyzes the circuit between zero and TSTART but does not write data to the output file. In the interval between TSTART and TSTOP, PSpice analyzes the circuit and writes the data to the output file.

The abbreviation UIC stands for **U**se **I**nitial **C**onditions. It is an optional keyword that tells PSpice not to determine a quiescent operating point before beginning the transient analysis. If the keyword UIC appears, PSpice uses the values specified in the element statements, that is, the IC = values. This option is of particular interest here, because our circuits do not contain nonlinear devices for which the quiescent operating point must be found.

7.3 TIME-RESPONSE OUTPUT

You may use any of three different ways to examine the output of a transient analysis. First, you may use the print statement to obtain a list of any desired voltage and current values for each time step specified in the transient analysis statement. Second, you may use the plot statement to generate a line printer plot of voltages and currents versus time. Finally, you may use the PROBE graphics postprocessor to generate high-quality graphs of voltage and current.

PRINTING

The format for the print statement is similar to that used for printing the results of dc analysis:

.PRINT TRAN OV1 < OV2 OV3 . . . >

Again, there are three fields. The first is the .PRINT command. The second specifies the type of analysis that was performed to yield the output variables—in this case, TRAN for transient analysis. The third consists of a list of output variables, which can either be voltages at nodes or currents through voltage sources. (See Chapter 4 to review the format used for voltages and currents.)

LINE PRINTER PLOTTING

You may use the plot statement to generate line printer plots on any printer. The format is:

.PLOT TRAN OV1 <(PLO1, PHI1)> < OV2

\+ <(PLO2, PHI2)> . . . >

The following notes describe the details of the plot statement.[†]

1. The .PLOT command statement defines the contents of one plot with one to eight variables—that is, OV1, OV2, . . . , OV8.

2. The syntax for the output variables is identical with that of the .PRINT command.

3. The low and high plot limits, i.e., PLO and PHI, are optional and may be specified after any of the output variables. All output

[†] A. Vladimirescu, K. Zhang, A. R. Newton, D. O. Pederson, and A. Sangiovanni-Vincentelli, *SPICE Version 2G User's Guide*, University of California at Berkeley, unpublished.

variables to the left of a pair of plot limits (PLO, PHI) will be plotted using the same lower and upper plot bounds.

4. If plot limits are not specified, PSpice will automatically determine minimum and maximum values of all output variables being plotted and scale the plot to fit. More than one scale will be used if the output variable values differ by orders of magnitude. This means that mixing such output variables on the same plot still yields a readable plot.

5. The overlap of two or more traces on any plot is indicated by the letter X.

6. When more than one output variable appears on the same plot, the first variable specified will be printed as well as plotted. If a printout of all variables is desired, a companion `.PRINT` command is required.

7. There is no limit on the number of `.PLOT` commands specified for each type of analysis.

PROBE PLOTS

PROBE is a software package that allows you to examine the results of PSpice circuit simulation using high-resolution graphics, both at the terminal and as hard copy. PROBE is simple to use, and it can be a very powerful tool in visualizing circuit analysis results. Before running PROBE you must be certain that a file named PSPICE.DEV exists on your PSpice directory. This file describes the devices available on your system. You must input the type of graphics display and printer you have in order for PROBE to generate the appropriate graphics. You may easily create or modify the PSPICE.DEV file from the PSpice Control Shell. (See Appendix A for more information.) If you are using PSpice and PROBE at the operating system level, you can create or modify the PSPICE.DEV file in any editor.

The PROBE software reads a data file that contains the results from PSpice analysis in a form especially suited to PROBE. You create this data file by using the command `.PROBE` in your PSpice source file. After completing a PSpice analysis, you may look at your results using PROBE, which you access from the Control Shell (details are presented in Appendix A) or directly from the operating system level by typing PROBE. PROBE searches for this special data file (usually called PROBE.DAT) and, upon finding it, displays a blank coordinate system on the screen with a set of menu selections at the bottom.

Select the **Add trace** menu item to display any of the voltages or currents in your circuit. The F4 function key gives you a list of available voltages and currents, in case you've forgotten the node numbers and device names used. You may select one or more for your display. The remaining menu items permit you to erase traces, reconfigure the x and y axes, add or delete other plots to the display, get a hard copy of the display, zoom in on your plot, move a cursor over your plot to get numerical values at any point, and label your plot. We illustrate many of these features throughout the rest of this manual. You should take some time to play with PROBE, as that is perhaps the best way to learn about all of its capabilities.

7.4 NATURAL RESPONSE

To find the natural response of a circuit you first find the initial inductor currents and capacitor voltages. When you know these values you can describe the circuit for transient analysis. In constructing the source file for a transient solution you must decide on the form of the output data, that is, whether to use a `.PRINT` command, a `.PLOT` command, a `.PROBE` command, or some combination of the three. Regardless of the form of the output data, you must choose appropriate values for TSTEP and TSTOP, the time increment and ending time for the PSpice analysis. For simple circuits, you can make decisions concerning TSTEP and TSTOP by means of a preliminary circuit analysis. Time constants and frequencies of oscillation are important characteristics that influence these decisions. For more complicated circuits, a trial-and-error approach may be necessary before you can satisfactorily choose a TSTEP and TSTOP.

Nilsson Computing the natural response of *RL* circuits (Section 8.1, pp. 254–262), *RC* circuits (Section 8.2, pp. 263–267), and *RLC* circuits (Secs. 9.1–9.2, pp. 322–339)

Example 9 illustrates how to use PSpice to find the natural response of a series *RLC* circuit. We incorporated a preliminary analytic solution as part of the example to permit checking the validity of the PSpice solution.

EXAMPLE 9

a) Analyze the natural response of the circuit shown in Fig. 27 with respect to the type of damping, the peak value of v_c, and the frequency of oscillation.

b) Write a PSpice source file to generate a plot of the voltage v_c versus t for the circuit shown in Fig. 27.

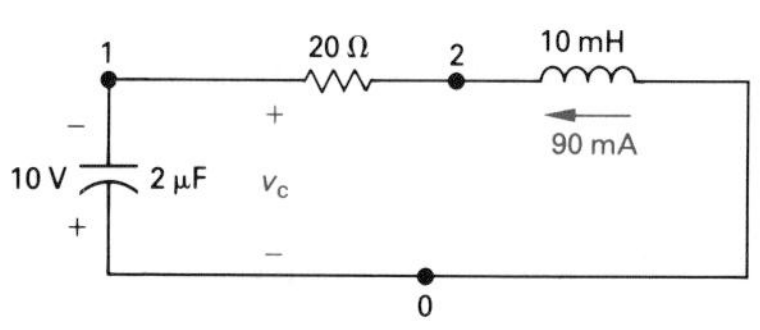

FIGURE 27 The circuit for Example 9.

c) Analyze the plot obtained in (b) and check its values against the predicted values obtained in (a).

d) Modify your source file to produce a PROBE data file. Then use PROBE to generate a plot of the voltage v_c versus t and use the PROBE cursor to identify the peak value of the voltage.

SOLUTION

a) For the series circuit shown in Fig. 27, $\alpha = R/2L = 1000$, $\alpha^2 = 10^6$, and $\omega_o^2 = 1/LC = 50 \times 10^6$. Hence the response is underdamped, and the roots of the characteristic equation are $s_1 = -1000 + j7000$ rad/s and $s_2 = -1000 - j7000$ rad/s. Therefore the form of the solution for v_c is

$$v_c = (B_1 \cos 7000t + B_2 \sin 7000t)e^{-1000t}.$$

From the initial conditions,

$$v_c(0) = -10 \qquad \text{and} \qquad \frac{dv_c(0)}{dt} = 45{,}000 \text{ V/s}.$$

Solving for B_1 and B_2 yields

$$v_c = (-10 \cos 7000t + 5 \sin 7000t)e^{-1000t} \text{ V}, \qquad t \geq 0.$$

Based on the solution for v_c, the peak value of v_c is 7.70 V, and it occurs at $t = 362.29$ μs. The damped frequency is 7000 rad/s, and thus the damped period is 897.60 μs.

b) From the preliminary analysis of the circuit we select TSTOP to be 2000 μs, so we can analyze the first two cycles of the response. Selecting TSTEP to be 20 μs gives 100 points between 0 and 2000 μs and should result in a reasonable graph. If the resulting plot is not to our liking, we can always rerun the analysis with different values of TSTEP and TSTOP. The PSpice source file is:

```
Natural Response of an RLC Circuit
C1      1 0     2E-6     IC = -10
R1      1 2     20
L1      2 0     10E-3    IC = -90e-3
.TRAN       20e-6     2000e-6     UIC
.PLOT       TRAN     V(1)
.PROBE
.END
```

Figure 28 shows the results of this PSpice analysis.

```
      TIME        V(1)
(*)----------    -1.0000E+01  -5.0000E+00    0.0000E+00   5.0000E+00    1.0000E+01

                     - - - - - - - - - - - - - - - - - - - - - - - - - - -
0.000E+00 -1.000E+01 *            .            .            .            .
2.000E-05 -9.015E+00 .  *         .            .            .            .
4.000E-05 -7.887E+00 .    *       .            .            .            .
6.000E-05 -6.664E+00 .        *   .            .            .            .
8.000E-05 -5.364E+00 .           *.            .            .            .
1.000E-04 -4.029E+00 .            .  *         .            .            .
1.200E-04 -2.652E+00 .            .     *      .            .            .
1.400E-04 -1.289E+00 .            .         *  .            .            .
1.600E-04  5.494E-02 .            .            *            .            .
1.800E-04  1.341E+00 .            .            .  *         .            .
2.000E-04  2.554E+00 .            .            .       *    .            .
2.200E-04  3.672E+00 .            .            .          * .            .
2.400E-04  4.671E+00 .            .            .           *.            .
2.600E-04  5.550E+00 .            .            .            .*           .
2.800E-04  6.276E+00 .            .            .            .   *        .
3.000E-04  6.867E+00 .            .            .            .     *      .
3.200E-04  7.286E+00 .            .            .            .      *     .
3.400E-04  7.567E+00 .            .            .            .       *    .
3.600E-04  7.670E+00 .            .            .            .       *    .
3.800E-04  7.642E+00 .            .            .            .       *    .
4.000E-04  7.446E+00 .            .            .            .      *     .
4.200E-04  7.135E+00 .            .            .            .      *     .
4.400E-04  6.677E+00 .            .            .            .    *       .
4.600E-04  6.128E+00 .            .            .            .   *        .
4.800E-04  5.463E+00 .            .            .            .*           .
5.000E-04  4.735E+00 .            .            .           *.            .
5.200E-04  3.927E+00 .            .            .         *  .            .
5.400E-04  3.089E+00 .            .            .       *    .            .
5.600E-04  2.211E+00 .            .            .     *      .            .
5.800E-04  1.334E+00 .            .            .  *         .            .
6.000E-04  4.580E-01 .            .            .*           .            .
6.200E-04 -3.887E-01 .            .           *.            .            .
6.400E-04 -1.198E+00 .            .         *  .            .            .
6.600E-04 -1.952E+00 .            .       *    .            .            .
6.800E-04 -2.638E+00 .            .     *      .            .            .
7.000E-04 -3.251E+00 .            .    *       .            .            .
7.200E-04 -3.770E+00 .            .  *         .            .            .
7.400E-04 -4.205E+00 .            . *          .            .            .
7.600E-04 -4.533E+00 .            .*           .            .            .
7.800E-04 -4.771E+00 .            .*           .            .            .
8.000E-04 -4.895E+00 .            *            .            .            .
8.200E-04 -4.934E+00 .            *            .            .            .
8.400E-04 -4.861E+00 .            *            .            .            .
8.600E-04 -4.712E+00 .            .*           .            .            .
8.800E-04 -4.463E+00 .            .*           .            .            .
9.000E-04 -4.151E+00 .            . *          .            .            .
9.200E-04 -3.758E+00 .            .  *         .            .            .
9.400E-04 -3.319E+00 .            .   *        .            .            .
```

(*continued*)

FIGURE 28 The PSpice-generated plot for Example 9.

```
9.600E-04 -2.822E+00 .            .     *       .            .             .
9.800E-04 -2.300E+00 .            .      *      .            .             .
1.000E-03 -1.744E+00 .            .       *     .            .             .
1.020E-03 -1.184E+00 .            .         *   .            .             .
1.040E-03 -6.162E-01 .            .          *  .            .             .
1.060E-03 -6.253E-02 .            .             *            .             .
1.080E-03  4.734E-01 .            .             .*           .             .
1.100E-03  9.788E-01 .            .             .  *         .             .
1.120E-03  1.445E+00 .            .             .   *        .             .
1.140E-03  1.868E+00 .            .             .    *       .             .
1.160E-03  2.234E+00 .            .             .     *      .             .
1.180E-03  2.548E+00 .            .             .      *     .             .
1.200E-03  2.795E+00 .            .             .      *     .             .
1.220E-03  2.986E+00 .            .             .       *    .             .
1.240E-03  3.103E+00 .            .             .       *    .             .
1.260E-03  3.164E+00 .            .             .       *    .             .
1.280E-03  3.153E+00 .            .             .       *    .             .
1.300E-03  3.091E+00 .            .             .       *    .             .
1.320E-03  2.962E+00 .            .             .       *    .             .
1.340E-03  2.790E+00 .            .             .      *     .             .
1.360E-03  2.561E+00 .            .             .      *     .             .
1.380E-03  2.301E+00 .            .             .     *      .             .
1.400E-03  1.998E+00 .            .             .    *       .             .
1.420E-03  1.675E+00 .            .             .   *        .             .
1.440E-03  1.326E+00 .            .             .  *         .             .
1.460E-03  9.698E-01 .            .             .  *         .             .
1.480E-03  6.047E-01 .            .             . *          .             .
1.500E-03  2.449E-01 .            .             .*           .             .
1.520E-03 -1.078E-01 .            .             *            .             .
1.540E-03 -4.438E-01 .            .            *.            .             .
1.560E-03 -7.583E-01 .            .           * .            .             .
1.580E-03 -1.047E+00 .            .          *  .            .             .
1.600E-03 -1.302E+00 .            .          *  .            .             .
1.620E-03 -1.526E+00 .            .         *   .            .             .
1.640E-03 -1.707E+00 .            .         *   .            .             .
1.660E-03 -1.853E+00 .            .        *    .            .             .
1.680E-03 -1.953E+00 .            .        *    .            .             .
1.700E-03 -2.016E+00 .            .        *    .            .             .
1.720E-03 -2.032E+00 .            .        *    .            .             .
1.740E-03 -2.014E+00 .            .        *    .            .             .
1.760E-03 -1.952E+00 .            .        *    .            .             .
1.780E-03 -1.861E+00 .            .        *    .            .             .
1.800E-03 -1.731E+00 .            .        *    .            .             .
1.820E-03 -1.579E+00 .            .         *   .            .             .
1.840E-03 -1.396E+00 .            .         *   .            .             .
1.860E-03 -1.198E+00 .            .          *  .            .             .
1.880E-03 -9.808E-01 .            .          *  .            .             .
1.900E-03 -7.564E-01 .            .           * .            .             .
1.920E-03 -5.229E-01 .            .            *.            .             .
1.940E-03 -2.907E-01 .            .            *.            .             .
1.960E-03 -6.014E-02 .            .             *            .             .
1.980E-03  1.638E-01 .            .             *            .             .
2.000E-03  3.789E-01 .            .             .*           .             .
                     - - - - - - - - - - - - - - - - - - - - - - - - - - - -
```

FIGURE 28 (*continued*)

c) From the output data, v_c(max) = 7.670 V, which occurs at t = 360 μs. The damped period is 1260 − 360, or 900 μs. All these measured values agree with our preliminary analysis. Note that the PSpice-generated line printer plot uses "E" in the exponential form for all numeric values.

d) Figure 29 presents the output from the PROBE graphics post-processor, along with the cursor output value for the peak voltage level. To create this plot, modify the source file by adding the . PROBE statement. Run PSpice to analyze the circuit again, and then run PROBE, either from the PSpice Control Shell or from the operating system. When the blank graph appears, select **Add trace** from the main menu and input V(1) as the variable. The plot will appear on the screen. Select the **Label** menu item and the **Text** submenu item to label the y axis as shown in the figure. Then select **Cursor** from the main menu and use the arrow keys to position the cursor at the maximum value of V(1). Watch the cursor values box in the lower right corner of the display. This box contains the x and y axis values of the first cursor, C1; the x and y axis values of the second cursor, C2 (which is positioned by moving the arrow keys while pressing the shift key); and the difference between the x axis values of the two cursors and the y axis values of the two cursors. PSpice and PROBE use the same characters for scale factor symbols (see Table 1) in x and y axis values and cursor box values, as you see in Fig. 29. Finally, you can make a hard copy of the display by selecting **Hard copy** from the **Cursor** menu.

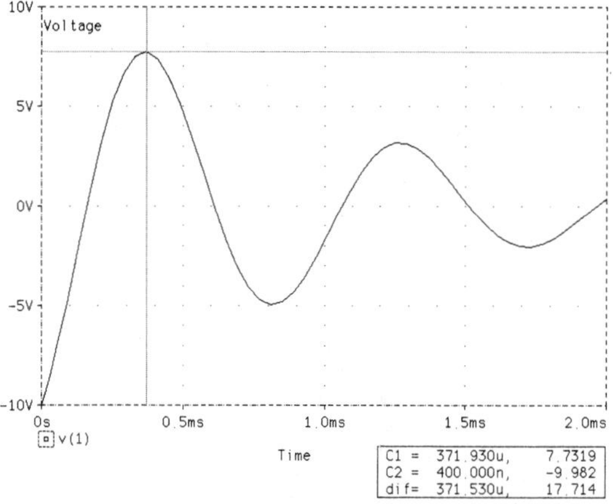

FIGURE 29 The PROBE-generated plot for Example 9.

PSpice also provides an alternative method for specifying initial conditions. This alternative method is most useful in circuits containing capacitors and semiconductor devices. If the circuit contains inductors carrying initial currents, you must specify the initial currents in the element statements. Instead of including the initial capacitor voltages as part of the element statements, you insert them by means of an . IC control statement in the source file.

The format for the . IC control statement is:

```
.IC    V(nodnum) = value    V(nodnum) = value . . .
```

Note that the node voltages are specified with respect to the reference node. Sometimes this method of entering initial capacitor voltages is more convenient than entering them on the individual element statements. In either case we continue to use the UIC keyword on the .TRAN control statement, because we do not need a preliminary bias analysis.

To illustrate use of the .IC control statement we rewrite the source file for Example 9 as:

```
Natural Response of an RLC Circuit
* Use of an .IC Statement
C1      1  0      2e-6
R1      1  2      20
L1      2  0      10e-3
.IC      V(1) = -10
.TRAN     20e-6     2000e-6     UIC
.PLOT     TRAN     V(1)
.PROBE
.END
```

We leave to you confirmation that this source file generates the same solution as the source file given in Example 9.

CHAPTER 8

SWITCHES

In this section we present two of several ways to model switches in PSpice. The first involves a specially constructed voltage source, and the second involves control by a circuit voltage or current.

Nilsson Sequential switching incorporates two or more switches in a single circuit: Section 8.5, pp. 282–286

8.1 PIECEWISE LINEAR SOURCE

The piecewise linear (PWL) time-dependent source allows you to define a single-valued function at a series of discrete times. PSpice determines the value of the source at intermediate values of time by using linear interpolation on the input values. The fact that the function must be single valued implies that the values of time used in the description of $f(t)$ must continually increase. One of the many uses for this type of source is as a model for a switch.

The syntax for a piecewise linear voltage source named Vgpwl, positive at N1 and connected at nodes N1 and N2, is:

```
Vgpwl    N1    N2    PWL(T1   V1   T2   V2   ...)
```

Each pair of values Tj, Vj (where j = 1, 2, . . .) specifies that the value of the source is Vj volts at Tj seconds. Note that you specify the times in increasing order so that T1 < T2 < T3 < . . . .

Example 10 illustrates the use of the piecewise linear source for modeling a transient analysis problem. This example also demonstates the .OP control statement discussed previously.

EXAMPLE 10

The circuit shown in Fig. 30 has been in operation for a long time. At $t = 0$ the 80-V source drops instantaneously to 20 V. Write a PSpice source file, using a PWL source, to model this circuit. The results of the analysis should be a plot of $v_o(t)$ versus t.

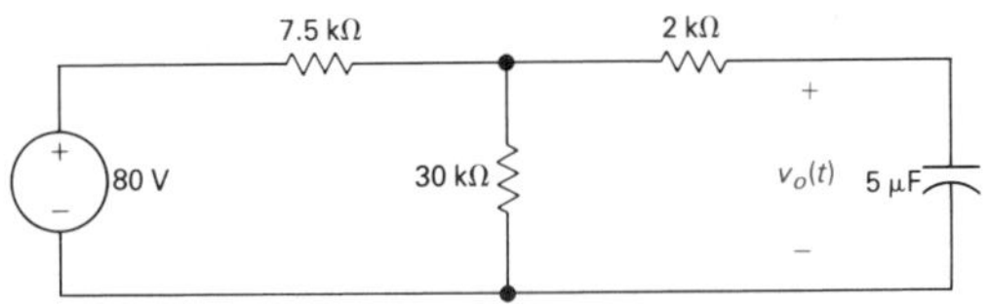

FIGURE 30 The circuit for Example 10.

SOLUTION

To simulate the drop in voltage from 80 to 20 V we insert a piecewise linear voltage source of opposite polarity in series with the 80-V dc source. We then step the piecewise source from 0 to 60 V over a short time interval. At the same time we make the duration of the 60-V step long enough that the circuit reaches its new steady-state condition. We must now decide how rapidly the source must rise to 60 V and how long it should remain at 60 V. We leave to you verification that the time constant for the circuit shown in Fig. 30 is 40 ms. We select a rise time for the piecewise linear voltage source of 40 ms/1000, or 40 μs. The duration of the 60-V source should be long compared with 40 ms, so we arbitrarily select 10 time constants, or 400 ms. The piecewise linear voltage source has the waveform shown in Fig. 31.

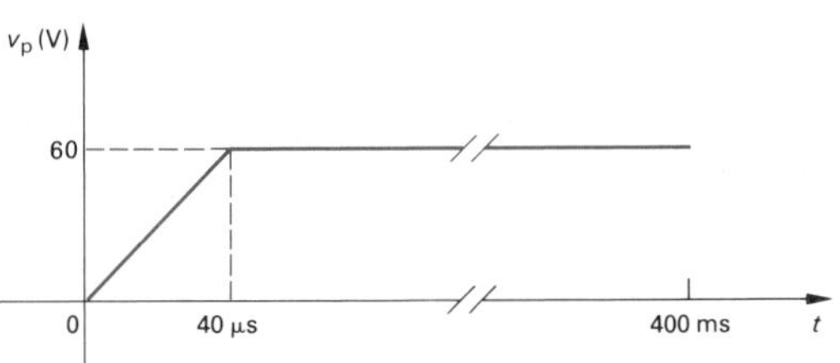

FIGURE 31 A piecewise linear voltage source.

Figure 32 contains the circuit shown in Fig. 30 redrawn in preparation for writing the PSpice source file. In describing the output for the analysis of this circuit as a plot of $v_o(t)$, which in this circuit is the voltage at node 4, V(4), we take advantage of the .OP control statement. Including the .OP control statement causes PSpice to calculate the initial conditions just prior to the application of the piecewise linear source. In other words, in this type of transient analysis problem we use PSpice to calculate initial conditions automatically. Thus no IC = value entries are required on the element statements, and we also omit the UIC keyword on the .TRAN control statement.

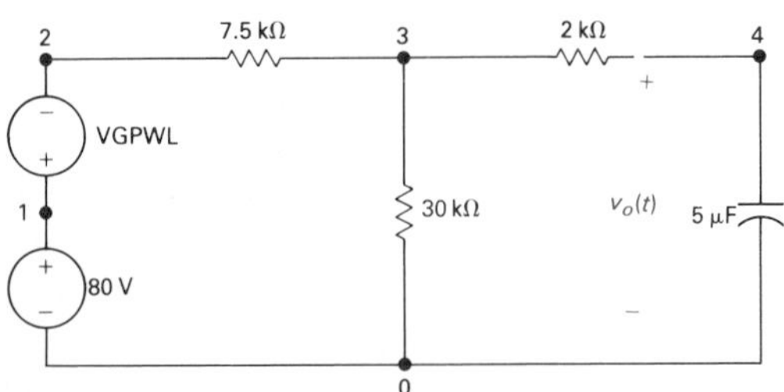

FIGURE 32 The circuit shown in Fig. 30 modified for PSpice analysis.

The PSpice source file for plotting V(4) is:

```
Modeling a Switch with a Piecewise Linear Source
V1      1  0     DC      80
Vgpwl   1  2     PWL(0   0   40e-6   60   400e-3   60)
R1      2  3     7.5e3
R2      3  0     30e3
R3      3  4     2e3
C1      4  0     5e-6
.OP
.TRAN      2e-3      200e-3
.PROBE
.END
```

We leave to you verification that the output of the PSpice analysis yields the following results.

t, ms	$v_o(t)$, V
0	64
40	33.67
80	22.50

Furthermore, the PSpice analysis shows $v_o(t)$ approaches 16 V as t approaches ∞.

8.2 VOLTAGE- AND CURRENT-CONTROLLED SWITCHES

PSpice provides two special circuit components that you may use to represent actual switches more accurately. One such component is the voltage-controlled switch, wherein the switch action depends on a voltage level elsewhere in the circuit. The other component is the current-controlled switch, wherein the switch action depends on the amount of current flowing through some other element in the circuit.

The formats for specifying the voltage-controlled and current-controlled switches are as follows:

S*xxx*	N1 N2	C1 C2	MNAME
Name of voltage-controlled switch	Node connections	Nodes for controlling voltage	Model name

W*xxx*	N1 N2	VNAME	MNAME
Name of current-controlled switch	Node connections	Name of zero-valued voltage source used to measure controlling current	Model name

Each format has four fields. The first field is the name of the switch: If it is voltage-controlled, the name must start with "S" (or "s"); if it is current-controlled, the name must start with "W" (or "w"). The second field identifies the branch of the circuit that contains the switch by naming both nodes. The third field identifies the *control* for the switch: For a voltage-controlled switch, the two nodes across which the controlling voltage is defined go in this field; for a current-controlled switch, the name of the zero-valued voltage source used to define the controlling current goes in this field. The fourth field contains the name of the model used to define the switch.

In addition to identifying the switch in a data statement, you must use the .MODEL statement to define the switch model. The .MODEL statement, which can be used for a variety of different devices in PSpice, permits you to define parameters that are particular to a given device. The form of the .MODEL statement is:

```
.MODEL    MNAME    DNAME(PVALUES)
```

where

MNAME is the name of the model, used in the fourth field of the data statement defining the switch;

DNAME is the name of the device being modeled, in this case VSWITCH for a voltage-controlled switch or ISWITCH for a current-controlled switch; and

PVALUES are the values of the parameters used to define the device.

Both voltage- and current-controlled switches have four device parameters. Each has a default value, so you need to set these parameter values in the .MODEL statement only if you desire a value other than the default value. The parameters are:

Parameter Name	Definition	Default value
RON	ON resistance, in ohms	1.0
ROFF	OFF resistance, in ohms	1+6
VON	control voltage level for ON, in volts	1.0
VOFF	control voltage level for OFF, in volts	0.0
ION	control current level for ON, in amps	1e–3
IOFF	control current level of OFF, in amps	0.0

Parameters RON, ROFF, VON, and VOFF are used for the VSWITCH model, and parameters RON, ROFF, ION and IOFF are used for the ISWITCH model.

For instance, the following statements identify a voltage-controlled switch named Sab, located between nodes 3 and 4 and controlled by a voltage drop between nodes 1 and 5. The model of the switch overrides the default values for the parameters RON and VON.

```
Sab    3  4    1  5    SMOD
.MODEL    SMOD    VSWITCH(RON = 5, VON = .2)
```

We illustrate the use of these switches in Example 12 in Chapter 9.

CHAPTER 9

STEP RESPONSE

The easiest way to find the step response of a circuit is to insert a dc source at the point where the step change takes place. To illustrate the insertion of a dc source we return to the circuit used in Example 9 (Fig. 27) and define the problem in Example 11.

Nilsson Computing the step response of *RL* and *RC* circuits (Section 8.3, pp. 267–275) and *RLC* circuits (Secs. 9.3 and 9.4, pp. 339–352)

EXAMPLE 11

a) The switch in the circuit shown in Fig. 33 is closed at $t = 0$. At the instant the switch is closed, the initial current in the 10-mH inductor is 90 mA and the voltage across the capacitor is 10 V. Figure 33 also shows the references for these initial conditions. Make a preliminary analysis of the step response and calculate the maximum value of v_c and the time at which it occurs.

b) Write a PSpice source file to plot v_c versus t from 0 to 3000 μs in steps of 20 μs. Use PROBE to generate the plot.

c) Compare the PSpice solution with the preliminary analysis.

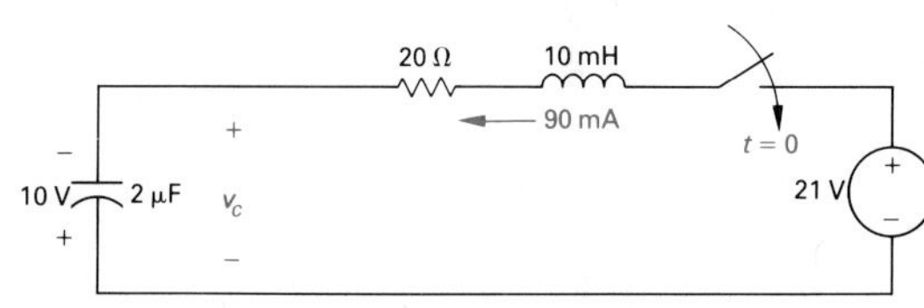

FIGURE 33 The circuit for Example 11.

SOLUTION

a) From the solution of Example 9 we already know that the response is underdamped. Furthermore, we know that $s_1 = -1000 + j7000$ rad/s, that $s_2 = -1000 - j7000$ rad/s, and that the damped period is 897.60 μs. The step-response solution for v_c takes the form:

$$v_c = v_f + (B_1' \cos 7000t + B_2' \sin 7000t)e^{-1000t}.$$

The initial values of v_c and dv_c/dt are the same as in Example 9, and the final value is 21 V. Hence $B_1' = -31$ V and $B_2' = 2$ V, so

$$v_c = 21 - (31 \cos 7000t - 2 \sin 7000t)e^{-1000t}, \qquad t \geq 0.$$

The maximum value of v_c is 41.22 V at 419.32 μs.

b) The PSpice source file is:

```
Step Response with an Ideal Switch
C1     0  1     2e-6     IC = 10
R1     1  2     20
L1     3  2     10e-3    IC = 90e-3
V1     3  0     DC      21
.TRAN 20e-6     3000e-6      UIC
.PROBE
.END
```

Figure 34 shows the plot generated by PROBE. We selected the **Cursor** menu item to enable the cursor to move. The arrow keys on your keyboard move the cursor labeled C1; hold down the shift key while pressing the arrow keys to move the cursor labeled C2. The cursors allow you to read the values of v_c(max) and T_d directly from the PROBE plot.

c) The output of the PSpice analysis yields

$$v_c(\text{max}) = 41.36 \text{ V}, \qquad \text{at } 436.8\ \mu\text{s};$$
$$T_d = 1336.8 - 436.8 = 900\ \mu\text{s};$$
$$v_f = 21 \text{ V}.$$

These results agree with the preliminary analysis.

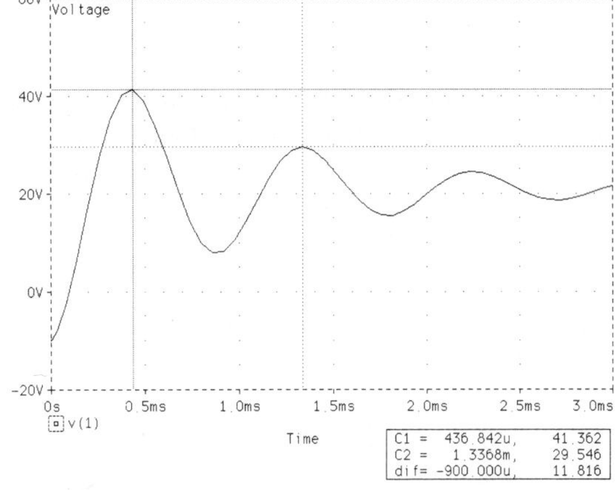

FIGURE 34 Plot of v_c versus t for Example 11.

Let's now consider a more realistic model of the step response: a modification of Fig. 33, wherein we added a switch controlled by the voltage step to the loop. Example 12 considers the effect of the nonzero switch resistance on the step respons. of the *RLC* circuit.

EXAMPLE 12

Use PSpice to model the circuit shown in Fig. 35. Assume that the voltage source is the same as that used in Example 11 and use the same values in the .TRAN statement. Assume further that the resistance of the switch in the ON position is 10 Ω. Then use PROBE to generate a plot of v_c versus t. Compare the results here with those in Example 11.

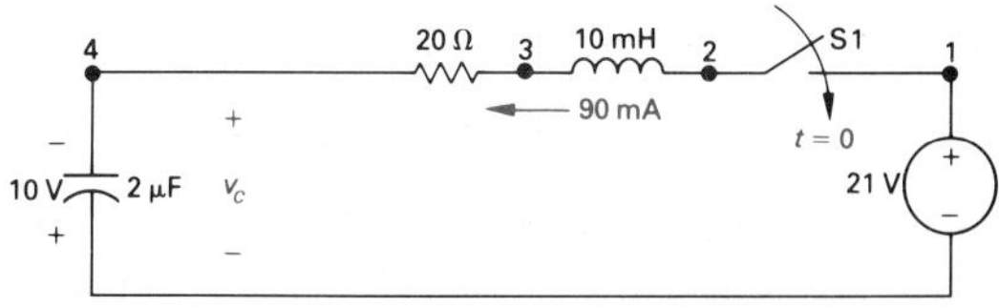

FIGURE 35 The circuit for Example 12.

SOLUTION

The PSpice source file that models the circuit in Fig. 35 is:

```
Step Response with a Realistic Switch
V1      1   0     PWL(0   0   1e-6   21   1000e-6   21)
S1      1   2     1   0       SMOD
L1      2   3     10e-3     IC = 90e-3
R1      3   4     20
C1      4   0     2e-6      IC = -10
.MODEL       SMOD       VSWITCH(RON = 10)
.TRAN      20e-6      3000e-6      UIC
.PROBE
.END
```

Figure 36 shows the PROBE plot of v_c versus t. Note that the values of v_c(max) and T_d are different in the two cases because of the resistance introduced here by the model of the switch.

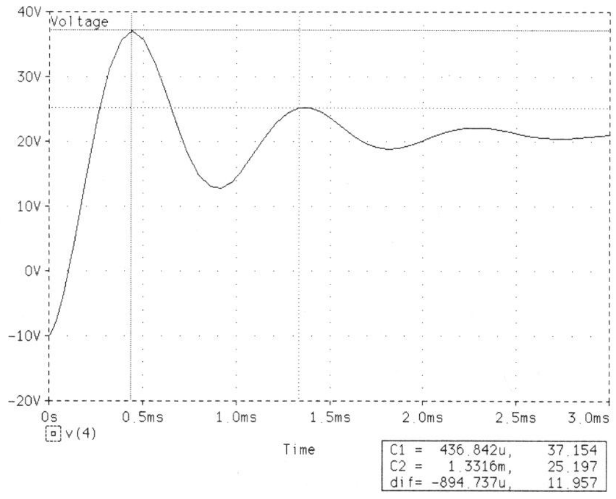

FIGURE 36 Plot of v_c versus t for the circuit of Example 12.

CHAPTER 10

VARYING COMPONENT VALUES

In PSpice you use the .STEP statement within a source file to vary the value of a circuit element over a specified range. This feature of computer-aided circuit analysis is very powerful, for it allows you to see the effects of changing a circuit parameter on the behavior of the circuit.

You may use the .STEP statement in one of two ways. The first allows sweeping the value over a range:

```
.STEP   STYPE   MNAME   SVALUE   EVALUE   CVALUE
```

where

STYPE is the sweep type—use LIN to sweep the value linearly from the starting to the ending value, use OCT to sweep the value logarithmically by octaves, and use DEC to sweep the value logarithmically by decades;

MNAME is the name of the component model defined in a .MODEL statement;

SVALUE is the starting value for the sweep;

EVALUE is the ending value for the sweep, which may be greater than or less than the starting value (that is, the sweep can go in either direction); and

CVALUE identifies the way the sweep value changes—if the sweep type is LIN, CVALUE is the increment used to compute the next value in the sweep, and if the sweep type is OCT or DEC, CVALUE is the number of values to generate starting at SVALUE and ending with EVALUE.

The other format for the . STEP statement defines a list of discrete values to use:

```
.STEP    MNAME    LIST    V1  V2  V3  ...
```

where MNAME again is the name of a component model defined in a . MODEL statement, and V1 V2 V3 ... are the values that PSpice will use for the component.

In order to vary the value of a resistor, capacitor, or inductor, you need to use a . MODEL statement to define a model for the component that you want to vary. We introduced the . MODEL statement in Chapter 8 with regard to voltage-controlled and current-controlled switches. The format is:

```
.MODEL    MNAME    DNAME(PVALUES)
```

where MNAME is the name of the model to be used in the . STEP statement, DNAME is a standard device name available in PSpice, and PVALUES are the values of the parameters that characterize the device you are modeling. The device names for resistors, inductors, and capacitors are RES, IND, and CAP, respectively. The parameters of interest that characterize resistors, inductors, and capacitors are the resistance multiplier R, the inductance multiplier L, and the capacitance multiplier C. You use the multipliers to compute the actual sweep value, which is the value specified in the . STEP statement multiplied by the multiplier. The default value for the multiplier is 1 for resistors, inductors, and capacitors.

For instance the PSpice statements

```
Ra      3  4     Rmod     1
.MODEL  Rmod     RES(R=1)
.STEP   RES      Rmod(R)  100  200  10
```

cause PSpice to analyze the circuit 11 times, once for each of the following values of Ra: 100 Ω, 110 Ω, . . . , 200 Ω. Note that computing the value for Ra involves multiplying the line value (the value at the end of the statement line that describes Ra) by the value of R in the . MODEL statement and then multiplying that by each succeeding value described in the . STEP statement.

Example 13 demonstrates how easily PSpice can analyze the same circuit several times with a different value of a given component used in each analysis. We used PROBE to illustrate graphically the varying behavior of the circuit as the component value changes.

EXAMPLE 13

Use the . STEP statement to vary the value of the resistor in Fig. 37 from 20 Ω to 100 Ω in 20-Ω steps. Then use PROBE to display the value of v_c versus t for each of the resistor values.

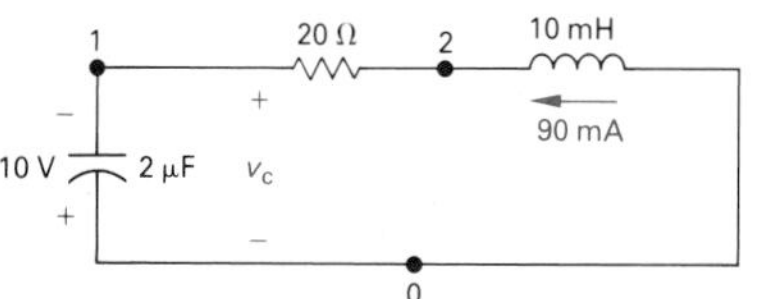

FIGURE 37 The circuit for Example 13.

SOLUTION

We modified the source file from Example 9 to include a varying resistor value by using the . MODEL and . STEP statements:

```
Effect of Varying R on RLC Natural Response
C1    1  0    2e-6
L2    2  0    10e-3    IC = -90e-3
R1    1  2    RMOD     1
.IC      V(1) = -10
.MODEL     RMOD      RES(R=1)
.STEP    LIN  RES   RMOD(R)      20, 100, 20
.TRAN     20e-6     2000e-6     UIC
.PROBE
.END
```

Figure 38 shows the PROBE plot of the capacitor voltage versus time. Note that the analysis using the smallest value of resistance, denoted □ on the graph, has the smallest damping ratio. Increasing the value of resistance increases the damping ratio. The largest value of resistance is denoted ▽ on the graph. We leave to you confirmation of these results analytically.

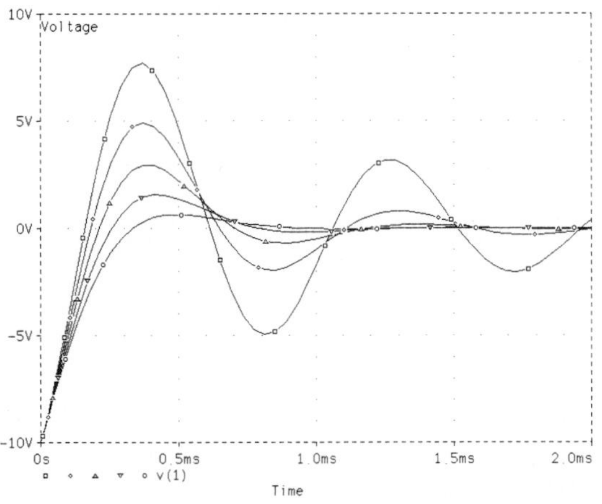

FIGURE 38 Plot of capacitor voltage versus time as the resistance is varied in the series *RLC* circuit of Fig. 37.

CHAPTER 11

SINUSOIDAL STEADY-STATE ANALYSIS

We now discuss the features of PSpice that permit you to examine how a circuit behaves in response to a sinusoidal input at a fixed frequency. This behavior is the ac steady-state response of the circuit.

11.1 SINUSOIDAL SOURCES

The following equations describe the damped sinusoidal voltage source:

$$v_g = \text{VO}, \qquad 0 \le t \le \text{TD};$$
$$v_g = \text{VO} + \text{VA}e^{-\theta(t-\text{TD})} \sin [2\pi\text{FREQ}(t - \text{TD}],$$
$$\text{TD} \le t \le \text{TSTOP};$$

where

VO = offset voltage in volts;

VA = amplitude in volts;

FREQ = frequency in hertz;

TD = delay in seconds; and

θ = damping factor in seconds^{-1}.

Nilsson Characteristics of sinusoidal sources: Section 10.1, pp. 372–377

The default values of TD and θ are zero, whereas FREQ defaults to 1/TSTOP. Figure 39 shows a graph of the damped sinusoidal voltage.

The syntax for a damped sinusoidal voltage source named VG, positive at node N1 and connected to nodes N1 and N2, is:

```
VG    N1  N2    SIN(VO  VA  FREQ  TD  THETA)
```

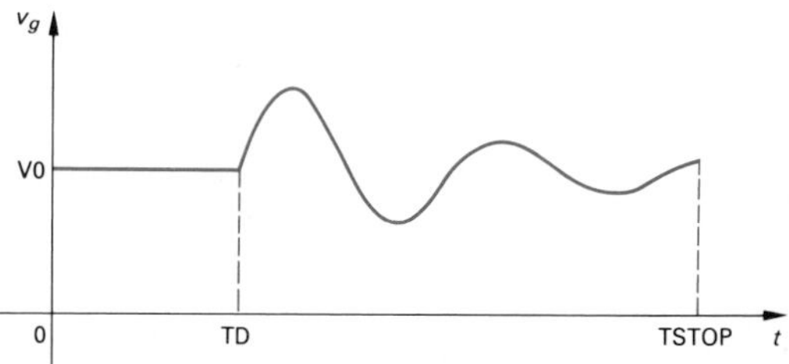

FIGURE 39 A damped sinusoidal voltage.

When constructing any of the time-dependent sources, you should check a plot of its waveform before using it in a specific problem. You can check the waveform by first applying it to a pure resistive circuit. The output waveform in a resistive circuit is a scaled replica of the input waveform. You may always choose a convenient scaling factor for the test circuit. For example, assume that you want to check the data statement for the damped sinusoidal voltage

$$v_g = 12.5 + 25e^{-0.1(t-1)} \sin [0.4\pi (t - 1)] \text{ V}.$$

If you apply this voltage to the voltage-divider circuit shown in Fig. 40, the output voltage is $0.8v_g$.

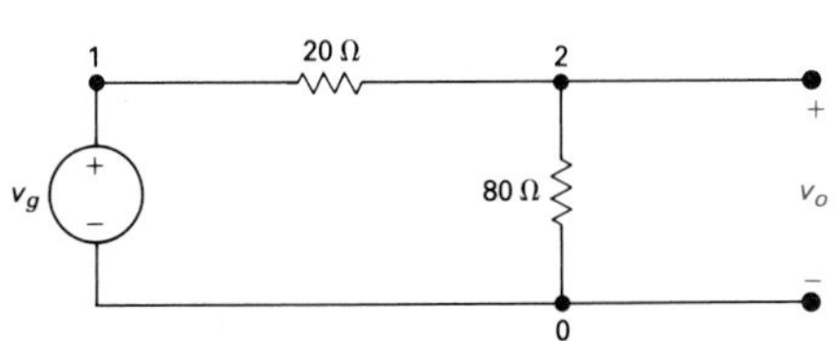

FIGURE 40 A voltage divider used to check v_g.

The following PSpice source file is used to generate a plot of v_o in PROBE. From that plot and tabulated values of v_o you can check the correctness of the v_g data statement.

```
Circuit Used to Check a Damped Sinusoid
Vg      1   0     SIN(12.5   25   0.2   1   0.1)
R1      1   2     20
R2      2   0     80
.TRAN     0.1      10       UIC
.PROBE
.END
```

11.2 SINUSOIDAL STEADY-STATE RESPONSE

Nilsson Steady-state circuit response to sinusoidal inputs: Secs. 10.2–10.9, pp. 377–406

The general format for the control statement that implements a sinusoidal steady-state analysis has five fields:

$$.\text{AC} \qquad \text{ATYPE} \qquad \text{NP} \qquad \text{FSTART} \qquad \text{FSTOP}$$

The first, . AC, identifies the type of analysis. The second, ATYPE, identifies the method used to vary the frequencies. For sinusoidal steady-state analysis you should use the keyword LIN. The third field, NP, contains the number of frequency points. The fourth, FSTART, is the starting frequency, and the fifth, FSTOP, is the final frequency. For sinusoidal steady-state analysis at a single frequency, set NP to 1 and make FSTOP equal to FSTART, where FSTART is the frequency of the sinusoidal sources. In steady-state sinusoidal analysis all sources in a circuit operate at the same frequency.

In addition to the .AC control statement, a .PRINT statement is required in order for you to get a printout of the desired voltages and currents. The output voltages and currents are phasor quantities, so you must specify whether you want polar or rectangular values. For voltages, follow the letter V with M (for magnitude), P (for phase), R (for real part), or I (for imaginary part). Thus the statement

```
.PRINT   AC   VM(1,3)   VP(1,3)
```

prints the magnitude and phase angle of the voltage between nodes 1 and 3, the reference polarity for V(1,3) being positive at node 1. The statement

```
.PRINT   AC   VR(1,3)   VI(1,3)
```

prints the real and imaginary components of V(1,3).

You may also generate current outputs in either polar or rectangular form. The statement

```
.PRINT   AC   IM(VSO1)   IP(VSO1)
```

prints the magnitude and phase angle of the current in voltage source VSO1. The current reference is through the source from the positive to the negative terminal. To obtain this same current in rectangular form, you use the statement

```
.PRINT   AC   IR(VSO1)   II(VSO1)
```

Example 14 illustrates how to use PSpice to find the single-frequency steady-state sinusoidal response.

EXAMPLE 14

The sinusoidal sources in the circuit shown in Fig. 41 are described by the equations

$$v_g = 20 \cos (10^5 t + 90°)\ \text{V} \quad \text{and} \quad i_g = 6 \cos 10^5 t\ \text{A}.$$

a) Use PSpice to find the magnitude and phase of v_o and i_o.

b) Check the PSpice results obtained in (a) against an analytic solution for V_o and I_o.

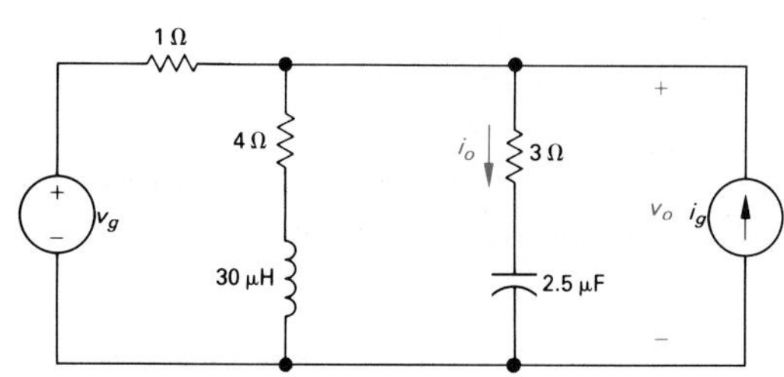

FIGURE 41 The circuit for Example 14.

SOLUTION

a) We redrew the circuit shown in Fig. 41 in Fig. 42 in preparation for writing the PSpice source file. The source file is:

```
AC Steady State Analysis
*  of the Circuit in Figure 42
V1     1  0     AC     20     90
I1     0  2     AC     6     0
R1     1  2     1
R2     2  3     4
R3     2  4     3
L1     3  0     30e-6
C1     4  5     2.5e-6
VIO    5  0     DC     0
.AC     LIN     1    15915.49    15915.49
.PRINT   AC   Vm(2)   Vp(2)    Im(VIO)   Ip(VIO)
.END
```

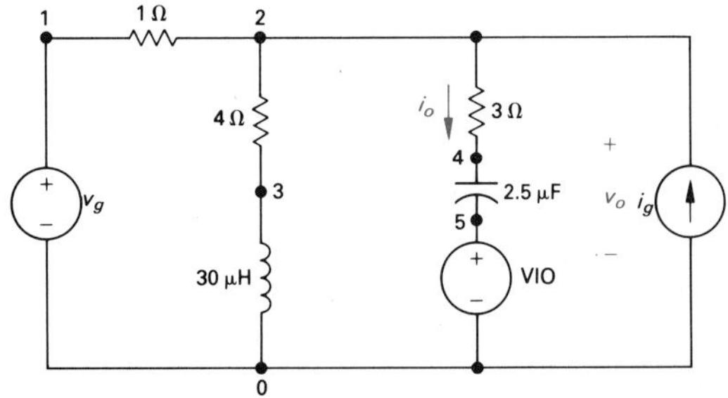

FIGURE 42 The circuit shown in Fig. 41 prepared for PSpice analysis.

Note that we must specify the frequency in a PSpice source file in hertz.

b) We obtain the solution for the phasor voltage V_o by solving a single node-voltage equation:

$$\frac{V_o}{4 + j3} + \frac{V_o}{3 - j4} + \frac{V_o - j20}{1} = 6\angle 0^\circ$$

The solution for V_o is

$$V_o = 16.31\angle 71.51^\circ \text{ V}.$$

Hence

$$I_o = \frac{V_o}{3 - j4} = 3.26\angle 124.64^\circ \text{ A}.$$

The pertinent output from the PSpice analysis is

$$VM(2) = 16.31 \quad \text{and} \quad VP(2) = 71.51^\circ.$$

Therefore $V_o = 16.31\angle 71.51^\circ$ V. Because

$$IM(VIO) = 3.261 \quad \text{and} \quad IP(VI) = 124.6^\circ,$$

$I_o = 3.261\angle 124.6^\circ$ A.

Example 15 illustrates sinusoidal steady-state circuit analysis when the circuit contains dependent sources.

EXAMPLE 15

The sinusoidal voltage source in the circuit shown in Fig. 43 is described by the relationship

$$v_g = 8.94 \cos(25000t - 116.57°) \text{ V}.$$

Use PSpice to find v_a and v_b in both polar and rectangular form.

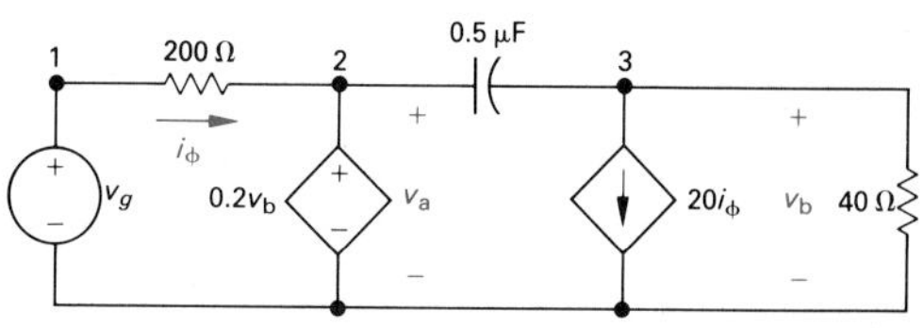

FIGURE 43 The circuit for Example 15.

SOLUTION

We use the following PSpice source file to obtain both the polar and rectangular expressions for V_a and V_b.

```
AC Steady State Analysis of the Circuit in Figure 43
Vg     1  0     AC      8.94     -116.57
E1     2  0     3  0      0.20
F1     3  0     Vg      -20
R1     1  2     200
R2     3  0     40
C1     2  3     0.5e-6
.AC      LIN    1     3978.87      3978.87
.PRINT     AC     Vm(2)     Vp(2)     Vr(2)     Vi(2)
.PRINT     AC     Vm(3)     Vp(3)     Vr(3)     Vi(3)
.END
```

In this source file the independent voltage source V1 measures the controlling current i_ϕ. Also, the current gain is -20, because the reference direction for i_ϕ is out of the positive terminal of v_g. Finally, the frequency of the source is converted from radians per second to hertz.

The pertinent output from the PSpice analysis is:

```
FREQ         VM(2)        VP(2)        VR(2)        VI(2)

 3.979E+03   1.599E+01   -4.928E-03   1.599E+01   -1.376E-03

FREQ         VM(3)        VP(3)        VR(3)        VI(3)

 3.979E+03   7.996E+01   -4.928E-03   7.996E+01   -6.878E-03
```

Then

$$V_a = 15.99\angle - 0.004928° = 15.99 - j0.001376 \text{ V}$$

and

$$V_b = 79.96\angle - 0.004928^\circ = 79.96 - j0.006878 \text{ V}.$$

We leave to you verification that these results are consistent with an analytic solution for the circuit.

CHAPTER 12

COMPUTING AC POWER WITH PROBE

Nilsson Defining and computing sinusoidal steady-state power quantities: Chapter 11, pp. 429–456

The interactive graphics capabilities provided by PROBE offer the opportunity to study the relationships among the various ac power quantities. In this chapter we investigate the relationships among current, voltage, instantaneous real power, average power, instantaneous reactive power, and reactive power for the circuit shown in Fig. 44. Although we introduce several new features of PROBE, you do not need any new PSpice constructs in order to investigate ac power.

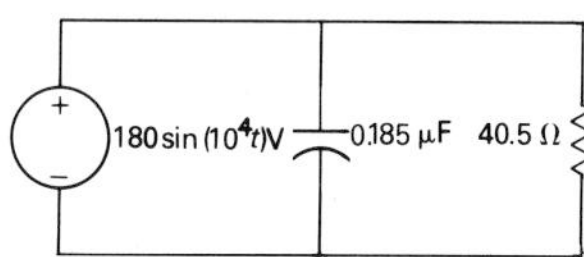

FIGURE 44 A circuit used to investigate AC power relationships.

To begin, construct a PSpice source file for the parallel *RC* circuit in Fig. 44, being sure to include the . PROBE statement:

```
Circuit Used to Demonstrate Analysis of AC Power
Vg     1  0    SIN(0   180   15915.49   0   0)
R1     1  0    40.5
C1     1  0    .185e-6
.TRAN     .2e-6     100e-6
.PROBE
.END
```

Have PSpice complete the analysis of this circuit and then invoke PROBE.

When the blank graph appears with the main menu below it, select the menu item **Plot control**. From the Plot Control menu select **Add plot** three times, which should give four blank plots on the screen. You can plot many different variables in these four plots and compare them. To switch among the four plots, you must choose **Select plot** from the Plot Control menu.

First, select the top plot on the screen—the symbol >>SEL should appear next to that plot. Exit the Plot Control menu, return to the Main menu, and select **Add trace**. Then choose the variable V(1) to plot the voltage supplied by the source to the

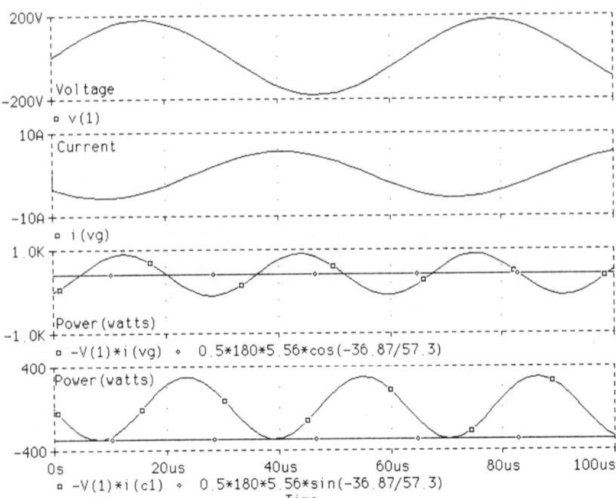

FIGURE 45 Some ac power characteristics from the circuit shown in Fig. 44: top plot, source voltage; second plot, source current; third plot, instantaneous and average power supplied by the source; bottom plot, instantaneous and reactive power absorbed by the capacitor.

parallel *RC* impedance. The plot should look like the first one (top) shown in Fig. 45.

Next, select the second plot on the screen and plot the current through the source. Note the phase relationship between the source voltage and current: The current leads the voltage because the load is capacitive. Compute the power factor angle of the load as follows:

$$Z_L = R \parallel 1/j\omega C$$

$$= \frac{R}{j\omega RC + 1};$$

$$\angle Z_L = 0 - \arctan(\omega RC)$$

$$= -\arctan(10^5 \cdot 40.5 \cdot 0.185 \times 10^{-6})$$

$$= -36.87°.$$

Now, select the third plot and graph the instantaneous power supplied by the source. Recall that the expression for instantaneous power is $v(t) \cdot i(t)$. PROBE allows you to supply algebraic expressions when it prompts for the variable to plot, so you simply type `-v(1)*i(Vg)`. Recall that `V(1)` is the voltage across the source (which you plotted in the top plot) and `i(Vg)` is the current through the source. Note that the minus sign (−) inverts the sign of the instantaneous power to conform to the PSpice sign convention. In this plot you have confirmed that the frequency of the instantaneous power is double the frequency of the voltage and current.

Now display the average power supplied by the source on top of the plot of the instantaneous power. The formula for average power is

$$P = \frac{1}{2} V_m I_m \cos(\angle Z_L).$$

Select the top two plots, one at a time, and use the cursor to find the maximum values $V_m = 180$ and $I_m = 5.56$. Select the third plot again and add the trace described by the expression:

```
.5*180*5.56*cos(-36.87/57.3)
```

Divide the impedance angle ($\angle Z_L = 36.87°$) by 57.3 to convert the angle from degrees to radians, as required by PROBE. In the third plot shown in Fig. 45, the equation for average power produced a straight line that runs through the center of the plot of instantaneous power, as you might have expected.

Use the fourth plot to examine the reactive power in the circuit. On the fourth plot generate a plot of the instantaneous power absorbed by the capacitor. Note that this is instantaneous *reactive* power, whose average value is zero. On top of this plot, place the plot of the reactive power supplied by the source, which is described by the expression:

```
.5*180*5.56*sin(-36.87/57.3)
```

You have completed the plot shown in Fig. 45. Now try to describe the relationship between the instantaneous and average reactive power absorbed by the capacitor.

This exploration of the ac power characteristics of the circuit shown in Fig. 44 using the power of PROBE necessarily has been brief. You should take the time to investigate further; try to construct plots of complex power, for example.

CHAPTER 13

MUTUAL INDUCTANCE

A PSpice source file requires three element statements to describe a pair of coils that are magnetically coupled. Each coil in such a pair requires a statement that contains the name, the polarity-sensitive node connections, and the self-inductance of the coil. The first-named node of each individual coil description is the *dotted* terminal. The third statement identifies the coupled coils by name and states the value of the coefficient of coupling k. The name used to identify the coupled coils must start with the letter K, and the value of k must be greater than 0 and less than 1.

Nilsson Analyzing circuits containing magnetically coupled coils: Secs. 13.2 and 13.3, pp. 506–514

We illustrate the PSpice description of two magnetically coupled coils with those shown in Fig. 46. If this pair of coupled coils is named KCCPAIR and the individual coils are L1 and L2, the three PSpice statements are:

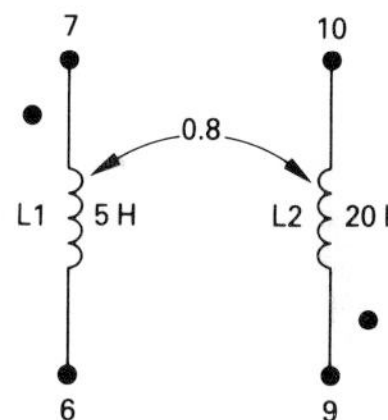

FIGURE 46 A pair of magnetically coupled coils.

```
L1        7  6   5
L2        9  10  20
KCCPAIR      L1  L2     0.8
```

Recall from our overview of data statements in Chapter 1 that every node in a circuit must have a dc path to the reference node. However, when a circuit contains magnetically coupled coils, one part of the circuit may be isolated from another part insofar as dc signals are concerned. If this happens, you must establish a dc path between the isolated portion of the circuit and the reference node. You may do so by either coalescing two isolated nodes into one node using a single conductor or by joining the two isolated nodes with a resistor. Figure 47 summarizes the two techniques.

Example 16 illustrates the PSpice analysis of a circuit containing magnetically coupled coils.

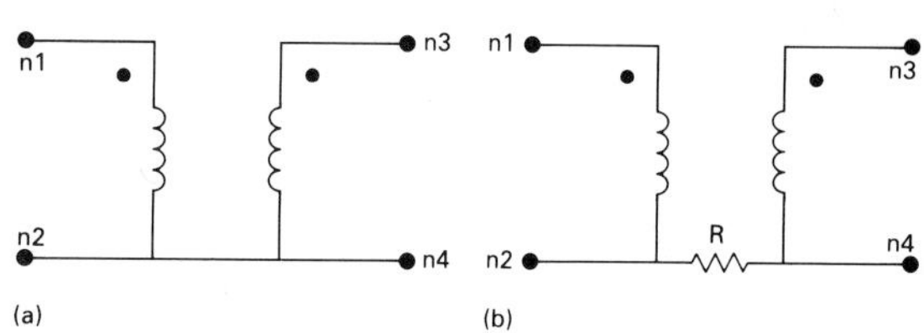

FIGURE 47 Providing a dc path between magnetically coupled coils: (a) coalescing isolated nodes with a single conductor; (b) joining isolated nodes with a resistor.

EXAMPLE 16

a) Use PSpice to find the rms amplitude and the phase angles of i_1 and i_2 in the circuit shown in Fig. 48 when $v_g = 141.42 \cos 10t$ V.

b) Check the PSpice solution by using it to show that the average power generated equals the average power dissipated.

FIGURE 48 The circuit for Example 16.

SOLUTION

a) Before redrawing the circuit for analysis by PSpice, we note that $\mathbf{V}_g = 100\angle 0°$ V rms, $f = 5/\pi \approx 1.59155$ Hz, and the coefficient of coupling $k = 6/\sqrt{64} = 0.75$. Figure 49 shows the PSpice circuit. The PSpice source file that yields the rms amplitude and the phase angle of i_1 and i_2 is :

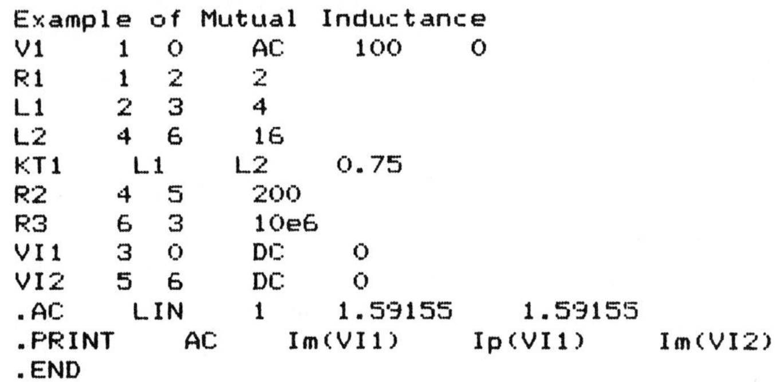

```
Example of Mutual Inductance
V1    1  0    AC    100    0
R1    1  2    2
L1    2  3    4
L2    4  6    16
KT1    L1   L2    0.75
R2    4  5    200
R3    6  3    10e6
VI1   3  0    DC    0
VI2   5  6    DC    0
.AC    LIN   1     1.59155     1.59155
.PRINT    AC     Im(VI1)     Ip(VI1)     Im(VI2)     Ip(VI2)
.END
```

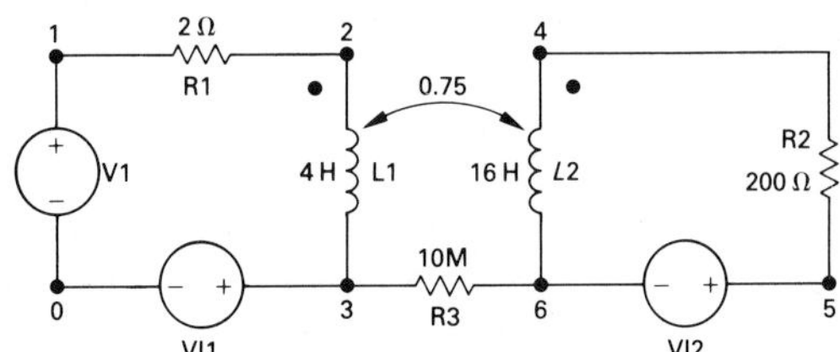

FIGURE 49 The circuit shown in Fig. 48 redrawn for PSpice analysis.

b) From the PSpice solution,

$$I_1 = 2.958\angle - 67.43° \text{ A(rms)} \quad \text{and}$$
$$I_2 = 0.6929\angle - 16.09° \text{ A(rms)}.$$

Using the PSpice solution, we find that the total average power generated is

$$P_g = (100)(2.958) \cos 67.43° = 113.532 \text{ W}.$$

The average power dissipated in R_1 is

$$P_1 = (2.958)^2(2) = 17.500 \text{ W}.$$

The average power dissipated in R_2 is

$$P_2 = (0.6929)^2(200) = 96.022 \text{ W}.$$

As $P_g = P_1 + P_2$, the PSpice solution is consistent with the conservation-of-energy principle.

CHAPTER 14

IDEAL TRANSFORMERS

Nilsson Analyzing circuits containing ideal transformers: Section 13.6, pp. 524–529

You can use a pair of magnetically coupled coils to model the ideal transformers in a PSpice source file by making L_1 and L_2 large enough that ωL_1 and ωL_2 are much greater than the other impedances in the circuit. In addition to setting large values for L_1 and L_2, setting the coefficient of coupling equal to 1 would be convenient. PSpice does not permit you to do so, but you may choose a value that is arbitrarily close to 1, say, 0.9999. You bring the turns ratio of the ideal transformer into the model by using the relationship $L_1/L_2 = (N_1/N_2)^2$.

Consider the circuit shown in Fig. 50, where $v_g = 50 \cos 1000t$ V. From the circuit diagram, $N_1/N_2 = 1/5$ and therefore $L_2 = 25L_1$. In picking values for L_1 and L_2, we want $\omega L_2 >> 200\ \Omega$. As $\omega = 1000$ rad/s, if we make $L_2 = 200$ H, ωL_2 will be three orders of magnitude greater than R_2. Therefore we pick $L_2 = 200$ H, and thus $L_1 = 8$ H.

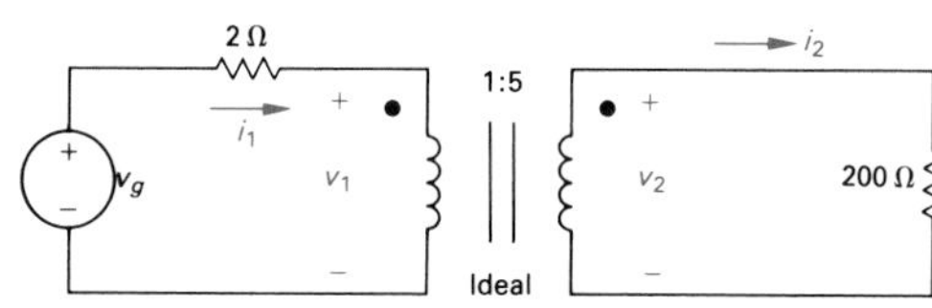

FIGURE 50 A circuit containing an ideal transformer.

We redrew the circuit in Fig. 50 in Fig. 51 to prepare for writing the PSpice source file to find the magnitudes and phase angles of v_1 and v_2. Note the insertion of a single conductor at the bottom of the circuit diagram. It establishes a dc path between all nodes and the reference node. Thus we reduced the five nodes in the circuit shown in Fig. 50 to four nodes in the circuit shown in Fig. 51.

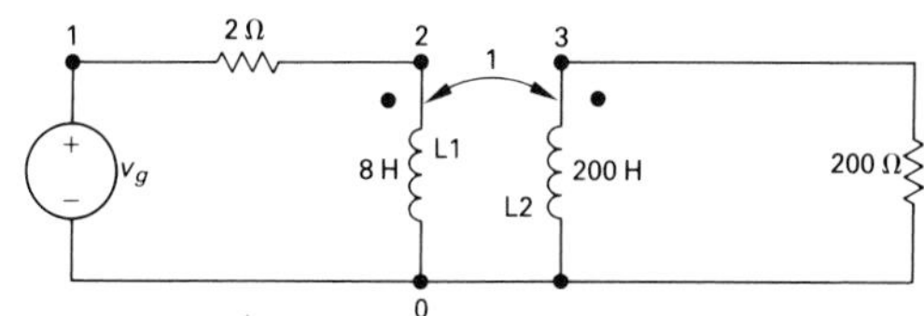

FIGURE 51 The circuit shown in Fig. 50 redrawn for PSpice analysis.

The PSpice source file that generates the magnitudes and phase angles of v_1 and v_2 is:

```
Example of Ideal Transformer Circuit
Vg     1  0     AC     50     0
R1     1  2     2
L1     2  0     8
L2     3  0     200
K1     L1    L2    .99999
R2     3  0     200
```

(continued)

where

LIN = linear variation;
NP = number of points;
DEC = decade variation;
ND = number of points per decade;
OCT = octave variation; and
NO = number of points per octave.

In each case FSTART and FSTOP are the starting and stopping frequencies in hertz.

If you select a linear variation, the frequency increment Δf is

$$\Delta f = \frac{\text{FSTOP} - \text{FSTART}}{\text{NP} - 1}$$

For example, if FSTOP = 1801 Hz, FSTART = 1 Hz and the number of points is 181. Then

$$\Delta f = \frac{1801 - 1}{181 - 1} = 10 \text{ Hz}$$

If you select a decade variation, the frequencies within a given decade are

$$f_k = \text{FSD} \cdot 10^{k/ND}$$

where FSD is the frequency at the start of the decade, ND is the number of points per decade, and k is an integer in the range from 1 to ND. For example, if ND = 20, FSTART = 50, and FSTOP = 5000, the frequencies in the decade between 50 and 500 would be

$$f_k = (50)10^{k/20},$$

where k = 1, 2, 3, . . . , 20. Hence

$$f_1 = (50)10^{1/20} = 56.10 \text{ Hz};$$
$$f_2 = (50)10^{2/20} = 62.95 \text{ Hz};$$
$$\vdots$$

If you select an octave variation, the frequencies within a given octave are

$$f_k = \text{FSO} \cdot 2^{k/NO}$$

where FSO is the frequency at the start of the octave, NO is the number of points per octave, and k is an integer in the range from 1 to NO. For example, if FSTART = 400 Hz, FSTOP = 3200 Hz, and NO = 10, the frequencies in the octave from 400 to 800 Hz would be

$$f_k = (400)2^{k/10}.$$

Hence

$$f_1 = (400)2^{0.1} = 428.71\text{Hz};$$
$$f_2 = (400)2^{0.2} = 459.48\text{ Hz};$$
$$\vdots$$

15.2 FREQUENCY RESPONSE OUTPUT

You may use the .PRINT, .PLOT, and .PROBE statements for PSpice output when the voltage and current variables are functions of frequency. The formats for .PRINT and .PLOT are similar to those used when outputting the results of dc analysis (see Chapter 4) and transient analysis (see Section 7.2):

```
.PRINT   AC   OV1   < OV2  OV3  . . . >
.PLOT    AC   OV1   < (PLO1, PHI1) >  < OV2 <(PLL2, PHI2) > . . . >
```

Note that the second field in each statement now contains AC to identify the type of analysis that produced the data to be printed or plotted.

In ac analysis you may specify the voltage and current variables in rectangular, polar, or decibel values. To obtain these values, you replace the V and I in the usual format used to specify voltage and current outputs with the following:

VR or IR = real part;
VI or II = imaginary part;
VM or IM = magnitude;
VP or IP = phase angle; and
VDB or IDB = $20 \log_{10}$ (magnitude).

For example, VR(N1, N2) yields the real component of the voltage between nodes N1 and N2, and IR(Vxxx) generates the real component of the current in the voltage source Vxxx.

Example 17 illustrates how to analyze the frequency response of a parallel RLC circuit with PSpice.

EXAMPLE 17

a) The current source in the circuit shown in Fig. 52 is 50 cos ωt mA. Use PROBE to plot v_o versus f from 1000 to 2000 Hz in increments of 10 Hz on a linear frequency scale.

b) From the PROBE plot, estimate the resonant frequency, the bandwidth, and the quality factor of the circuit.

c) Compare the results obtained in (b) with an analytic solution for f_o, Δf, and Q.

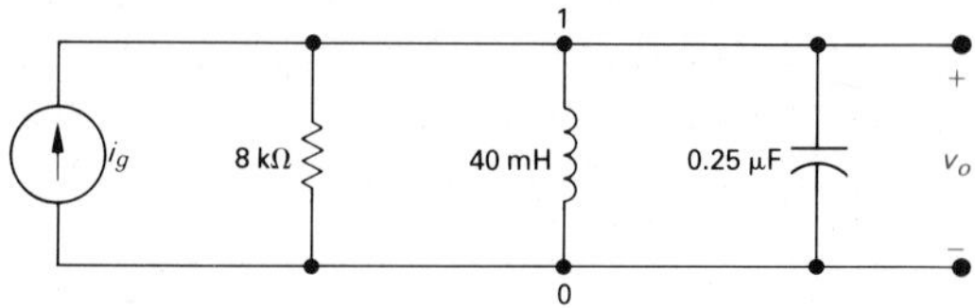

FIGURE 52 The circuit for Example 17.

SOLUTION

a) The PSpice source file for obtaining a plot of v_o versus f is:

```
Parallel RLC Circuit Used to
*  Illustrate Frequency Response
Ig      0  1      AC      50e-3     0
R1      1  0      8e3
L1      1  0      40e-3
C1      1  0      .25e-6
.AC     LIN       101        1000       2000
.PROBE
.END
```

b) Figure 53 shows the PROBE plot of v_o versus f. We need to select the **X Axis** menu item and choose a linear range such as shown in Fig. 53. Using the **Cursor** menu item, we note that the peak amplitude of 400.0 V occurs at a frequency of 1590.2 Hz in Fig. 53 (a). Thus we estimate the resonant frequency at 1590 Hz.

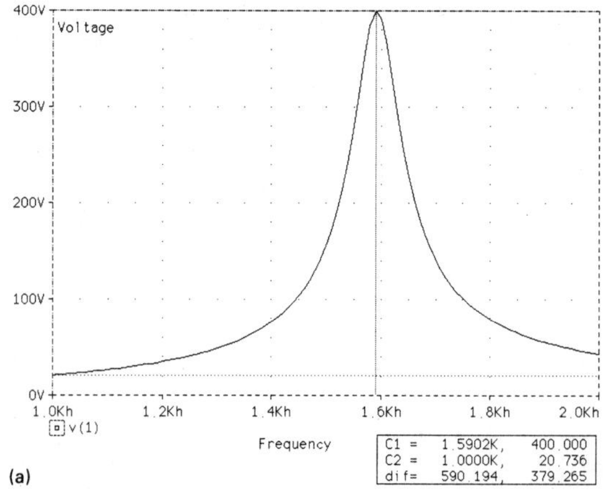

(a)

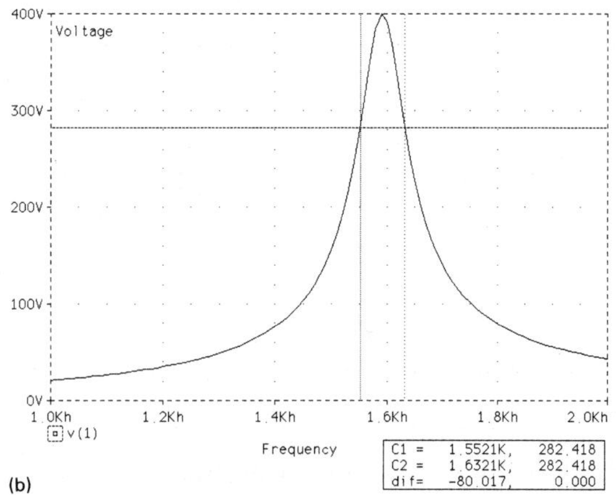

(b)

FIGURE 53 PROBE plot of v_o versus f for Example 17, using the cursor in (a) to identify the resonant frequency and in (b) to identify the bandwidth.

To estimate the bandwidth we use both cursors to find the frequencies where $v_o = 399.7/\sqrt{2} = 282.63$ V. From the PROBE plot in Fig. 53 (b), the closest values are 282.42 V at 1552 Hz and 282.42 V at 1632 Hz. Thus we estimate the bandwidth to be 1632 − 1552, or 80 Hz.

We calculate the quality factor from the relationship $Q = f_o/\Delta f = 1590/80 = 19.88$.

c) A direct analysis of the circuit yields

$$f_o = 1591.55 \text{ Hz};$$
$$f_1 = 1551.76 \text{ Hz};$$
$$f_2 = 1631.34 \text{ Hz};$$
$$\Delta f = f_2 = f_1 = 79.58 \text{ Hz};$$
$$Q = f_o/\Delta f = 20.$$

We summarize the comparison with the PSpice analysis as follows:

Quantity	**Analysis**	**PSpice**
f_o	1591.55	1590.20
f_1	1551.76	1552.10
f_2	1631.34	1632.10
Δf	79.58	80.00
Q	20.00	19.88

If necessary, we can obtain a more accurate analysis around the resonant frequency from PSpice by rerunning the program with different values of NP, FSTART and FSTOP.

In Example 18 we use the .STEP command to vary the value of the capacitor in the circuit shown in Fig. 52. We use PROBE to look at the effect of this variation on the frequency response characteristics of the circuit.

EXAMPLE 18

Modify the PSpice source file for Example 17 to step the capacitor values from 0.15 μF through 0.35 μF in increments of 0.5 μF. Then use PROBE to display the frequency response characteristics for all values of capacitance. Comment on the effect of the changing capacitance.

SOLUTION

The modified PSpice source file is:

```
Effects of Varying C on
* Parallel RLC Circuit Frequency Response
Ig      0  1     AC      50e-3     0
R1      1  0     8e3
L1      1  0     40e-3
C1      1  0     CMOD      1
.MODEL      CMOD      CAP(C=1)
.STEP  CAP   CMOD(C)  0.15e-6  .35e-6  .05e-6
.AC     LIN     101      500      2500
.PROBE
.END
```

Figure 54 shows the PROBE plot. The smallest value of capacitance produced the plot farthest to the right, denoted □. As the capacitance increases, the plots move to the left. Hence the resonant frequency *decreases* as the capacitance *increases*. But we expect this result because the equation for resonant frequency for an *RLC* circuit is

$$\omega_r = \sqrt{\frac{1}{LC}}.$$

Also, as the capacitance increases, the resonant peak becomes sharper. That is, as the capacitance increases, the quality becomes higher. This result, too, comes as no surprise, because the equation for Q in a parallel *RLC* circuit is

$$Q = R\sqrt{\frac{C}{L}}.$$

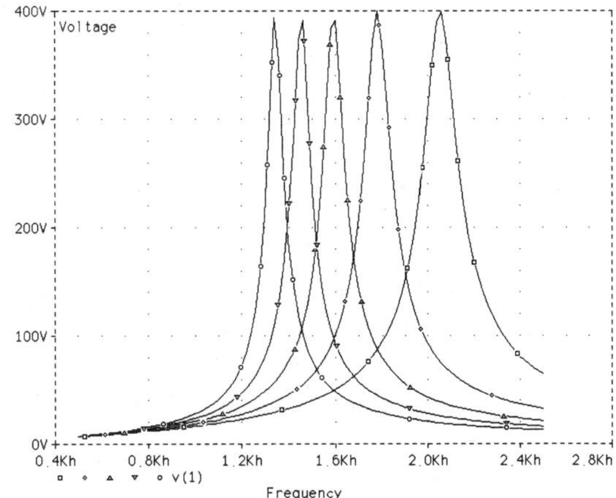

FIGURE 54 The relationship between changing capacitance and circuit frequency response.

15.3 BODE PLOTS WITH PROBE

Nilsson Generating Bode plots from a circuit's transfer function: Secs. 17.6 and 17.7, pp. 705–722

In Section 15.2 we described the method for generating a frequency response plot with PSpice and PROBE. The `.PRINT` statement for frequency response data can provide you with output voltages and currents in several different forms: real and imaginary parts, magnitude and phase angles, and dB magnitude and phase angles. PROBE also allows you to plot frequency response data by using these same forms. In Section 15.2 we plotted linear voltage magnitude versus linear frequency. We now turn to another form: dB voltage magnitude versus log frequency and phase angle versus log frequency. This is the form used to generate Bode plots. Example 19 compares the exact dB voltage magnitude versus log frequency and phase angle versus log frequency plots to the Bode straight-line approximations using PROBE.

EXAMPLE 19

Construct a PSpice source file to generate the frequency response of the circuit shown in Fig. 55 for three different values of resistance: 5 Ω, 50 Ω, and 500 Ω. Then use PROBE to plot the output voltage magnitude in dB and output voltage phase angle versus log frequency. Finally, use the **Label** menu to overlay a straight line Bode approximation plot and comment.

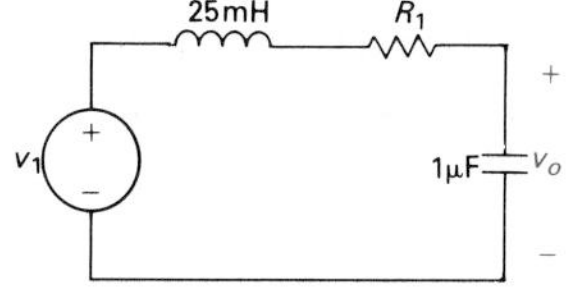

FIGURE 55 The circuit for Example 19.

SOLUTION

In order to analyze the circuit three times with three different values of resistance, we use a model for the resistor and step through the values. The values required are listed in the .STEP command. We need to perform some preliminary analysis of the circuit to generate the values used in the .AC command: the starting and ending frequencies. We choose them so that the frequency band of interest, that is, the frequency band that contains the natural frequency of the circuit, is centered between the two end-point frequencies.

Because this is a series *RLC* circuit, we know that the center frequency and the bandwidth are given by

$$\omega_o = \frac{1}{\sqrt{LC}} \quad \text{and} \quad \beta = \frac{R}{L}.$$

For the circuit shown in Fig. 55, the center frequency is

$$\omega_o = \frac{1}{\sqrt{(25 \times 10^{-3})(1 \times 10^{-6})}} = 6324.56 \text{ rad/sec}$$
$$\approx 1000 \text{ Hz},$$

and the bandwidth ranges from

$$\beta_{\text{narrow}} = \frac{5}{25 \times 10^{-3}} = 200 \text{ rad/sec}$$
$$\approx 32 \text{ Hz}$$

to

$$\beta_{\text{wide}} = \frac{500}{25 \times 10^{-3}} = 20000 \text{ rad/sec}$$

$$\approx 3.2 \text{ kHz.}$$

Thus we choose 10 Hz as the starting frequency and 100,000 Hz as the ending frequency, which places the center frequency in the middle and provides adequate room for the wideband circuit's response. The PSpice source file is:

```
RLC Circuit Used to Illustrate Bode Plots
C1      1  0     1e-6
R1      1  2     RMOD      1
L1      2  3     25e-3
V1      3  0     AC      1
.MODEL      RMOD      RES(R=1)
.STEP      RES      RMOD(R)      LIST    5, 50, 500
.AC     DEC      300      10      100000
.PROBE
.END
```

After completing the analysis we enter PROBE and use the menu system to create two plots on the screen. In the upper plot, we add the trace of VDB(1), which is the magnitude of the voltage across the capacitor in dB. Note that the frequency axis is already logarithmic, because the .AC command specified frequencies in decades. In the lower plot, we add the trace of VP(1), which is the phase angle of the voltage across the capacitor.

Now use the **Label** menu selection and the **Line** option to overlay the straight-line Bode magnitude and phase-angle plots. The transfer function for this circuit takes the form

$$H(s) = \frac{1/LC}{s^2 + (R/L)s + 1/LC}$$

and so has two poles and no zeros. The frequency of the two poles is the center frequency, ω_o, which we have already calculated as 1000 Hz. The straight-line phase angle plot consists of three lines: for $\omega < u_1$, the straight line has a value of 0° for all frequencies; for $\omega > u_2$, the straight line has a value of $-180°$ for all frequencies; and for $u_1 < \omega < u_2$, the straight line has a slope of $-132/\zeta$ degrees/decade. Note that $u_1 = 4.81^{-\zeta}\omega_o$, $u_2 = 4.81^{\zeta}\omega_o$ and ζ is the damping coefficient. For the straight-

Nilsson Constructing straightline Bode plot approximations: Secs. 17.6 and 17.7, pp. 706–709, 711–713, 715–716, and 718–719

line approximation constructed in Fig. 56, we use $\zeta = 0.7$, so $u_1 = 333$ Hz and $u_2 = 3002$ Hz. The two straight lines intersect at the natural frequency, ω_o. Use the **Line** cursor to plot these two lines. The straight-line phase angle plot consists of three lines: for $\omega < 0.1\omega_o$, the straight line has a value of 0° for all frequencies; for $\omega > 10\omega_o$, the straight line has a value of $-180°$ for all frequencies; and for $0.1\omega_o < \omega < 10\omega_o$, the straight line has a slope of $-90°$/decade. Use the **Line** cursor to plot these three lines.

Figure 56 presents the PROBE plot and the overlaid Bode plot. Note that, as expected, the accuracy of the Bode approximation is very good outside the band of frequencies, where the magnitude and phase angle are changing. Within this band, which spans one decade on either side of ω_o, the accuracy of the Bode approximation depends on the damping ratio of the underlying second-order circuit. When the damping ratio is small ($\zeta \ll 0.7$), the approximation is not particularly good. Similarly, when the damping ratio is large ($\zeta \gg 0.7$) the approximation is also not particularly good. Only when the damping ratio is close to 0.7 does the approximation more nearly represent the actual frequency response.

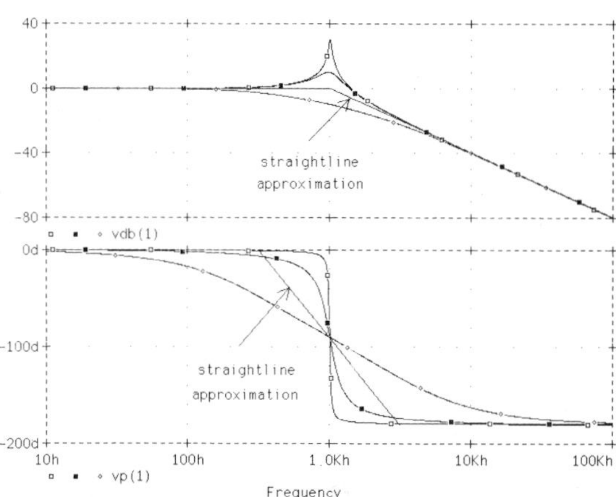

FIGURE 56 Actual frequency response and straight-line approximation of the frequency response for the circuit shown in Figure 55.

CHAPTER 16

PULSED SOURCES

The syntax for a periodic pulse voltage source named Vpp1, connected to nodes 1 and 0, and positive at node 1 is

```
Vpp1   1  0    PULSE(V1  V2  TD  TR  TF  PW  PER)
```

The pulse parameters are:

V1 = initial value (volts);

V2 = pulsed value (volts);

TD = delay time (seconds);

TR = rise time (seconds);

TF = fall time (seconds);

PW = pulse width (seconds);

PER = period (seconds).

The default values for the time parameters are TD = 0, TR = TSTEP, TF = TSTEP, PW = TSTOP, and PER = TSTOP. TSTEP and TSTOP are specified in the .TRAN control statement. Figure 57 shows a graph of the periodic pulse. We discuss an illustrative application of a periodic pulse source in Example 20.

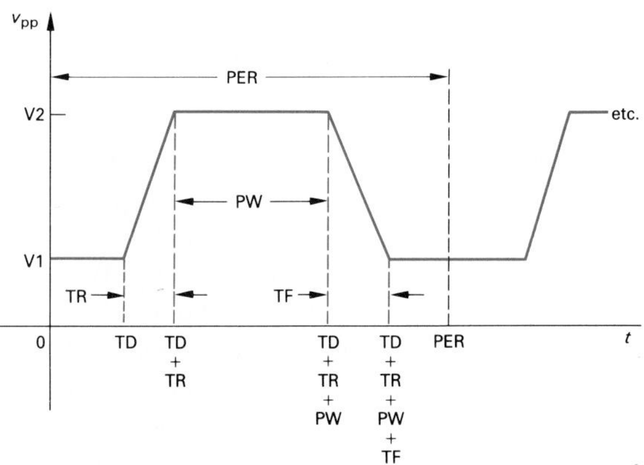

FIGURE 57 A periodic pulse voltage source.

EXAMPLE 20

The circuit shown in Fig. 58 is excited by an impulsive current source, where $i_s = 7\delta(t)\ A$. Use PSpice to plot $v_o(t)$ versus t.

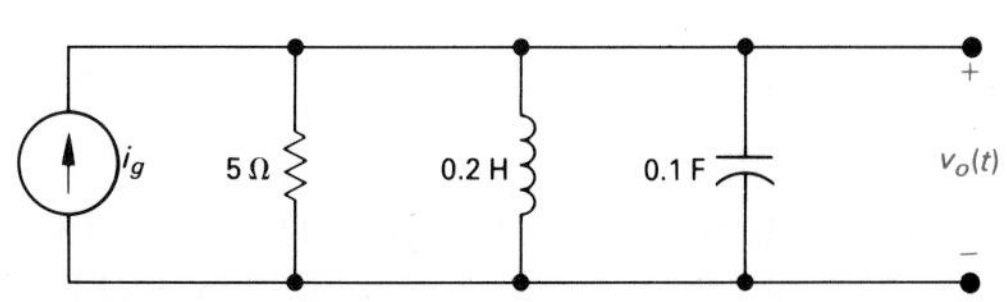

FIGURE 58 The circuit for Example 20.

SOLUTION

To use PSpice to obtain a plot of $v_o(t)$ we have to construct a current pulse that satisfactorily models the given impulsive current source. We begin by making a preliminary analysis of the circuit and studying the nature of the transient behavior. In particular, we need to get a feel for how rapidly the transient response decays. This understanding enables us to build a current pulse whose duration is brief compared to the life of the transient voltage. The roots of the characteristic equation contain the information we seek.

For the circuit shown in Fig. 58, the characteristic equation is

$$s^2 + \frac{1}{RC}s + \frac{1}{LC} = 0.$$

For the given values of R, L, and C, the roots are

$$s_{1,2} = -\alpha \pm j\omega_d = -1 \pm j7 \text{ rad/s}.$$

Hence the impulsive response takes the form

$$v_o(t) = B_1 e^{-\alpha t} \cos \omega_d t + B_2 e^{-\alpha t} \sin \omega_d t,$$

and therefore the duration of the pulse should be short compared to α, or 1 s. Furthermore, in order for the transient to die out before the next pulse arrives, the period of the pulse train should be on the order of 5 α or 5 s.

We select a pulse duration of $\alpha/1000$, or 1 ms. So long as the duration of the pulse is brief with respect to 1 s, the exact shape of the pulse is not critical. For convenience we construct the trapezoidal pulse shown in Fig. 59. We arbitrarily choose the break points to occur at $T_o/5$ and $4T_o/5$.

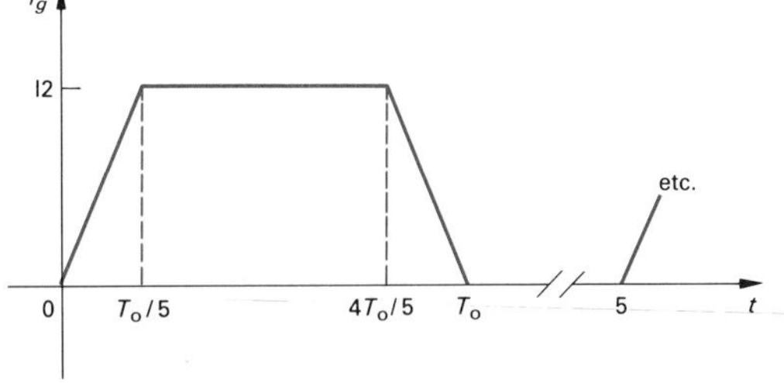

FIGURE 59 The current pulse for Example 20.

Having decided on the general shape of the current pulse, we must now determine the height of the pulse, I2. We know from the material on the impulse function that the impulsive current $7\delta(t)$ instantaneously charges the 0.1 F capacitor to 70 V. (When the impulse has passed, the response is identical to the natural response of the circuit, that is, the response associated with an initial capacitor voltage of 70 V.) Therefore the current pulse must charge the capacitor to 70 V during its duration of T_o seconds. Thus

$$v_o = \frac{1}{C}\int_0^{T_o} i \, dt = 70.$$

The integral of the current is simply the area under the current versus time curve, which we can easily calculate in this case without integrating:

$$v_o = \frac{4}{5}\frac{I2T_o}{C} = 70.$$

As $T_o = 1$ ms and $C = 0.1$ F, we have

$$I2 = 8750\ A.$$

If we label the top node in the circuit shown in Fig. 58 as 1, the PSpice source file that produces a plot of $v_o(t)$ versus t is:

```
Example of a Pulsed Source
Ig   0   1   PULSE(0   8750   0   0.2M   0.2M   0.6M   5)
R1   1   0   5
L1   1   0   0.2       IC = 0
C1   1   0   0.1       IC = 0
.TRAN     10e-3   1400e-3   0   1e-3     UIC
.PLOT     TRAN    V(1)
.END
```

Note in the .TRAN control statement in this source file that the plotting increment is 10 ms, that the computation increment is 1 ms, and that TSTOP is approximately $1.5T_d$, where $T_d = 2\pi/7$ s.

The output of the PSpice analysis gives the following results. The voltage at 1 ms is 69.24 V; the first minimum value of v_o occurs at 411 ms and is -47.13 V; and the damped period is $1301 - 411 = 890$ ms. We leave to you verification of these results.

CHAPTER 17

FOURIER SERIES

Both PSpice and PROBE can help you to understand the frequency spectrum characteristics of periodic waveforms. In PSpice the .FOUR command computes and prints the magnitude and phase angle of the Fourier series decomposition of the periodic signal at dc, the fundamental frequency, and the second through the ninth harmonic frequencies. Transient analysis must precede Fourier series analysis, because the Fourier series coefficients are computed for the last complete cycle for the transient analysis waveform. Therefore, the duration of the transient analysis must be a least one cycle of the waveform. If you want PSpice to compute the Fourier series coefficients of the steady-state waveform, then the transient analysis should be performed for a number of cycles sufficient to insure the decay of the natural response.

Nilsson Calculating the Fourier series representation of a periodic waveform: Secs. 18.1–18.4, pp. 738–752 and Section 18.8, pp. 764–767

The format for the .FOUR command is:

```
.FOUR    FFUND    OUTVAR1  <OUTVAR2  ...>
```

The first field contains the .FOUR command. The second field contains the fundamental frequency of the periodic signal in Hertz; recall that the fundamental frequency in Hertz is the inverse of the period of the signal in seconds. The third field contains the variables for which the Fourier coefficients are computed. As usual, these variables must be voltages defined between two nodes or a current defined through a voltage source. Including a .PRINT or a .PLOT command is not necessary, as the Fourier coefficients automatically print in the output file as PSpice analysis proceeds.

If a PSpice source file contains a .TRAN statement, you may use PROBE to calculate and display the Fourier transform coefficients whether or not the source contains a .FOUR command. In fact, PROBE calculates the Fourier transform

Nilsson Defining the Fourier transform and its properties: Secs. 19.1–19.6, pp. 789–805

The PSpice source file that describes the circuit shown in Fig. 61 with the smaller capacitor value and the input shown in Fig. 60 is:

```
Fourier Series Analysis of RC Circuit with Pulse Input -- Small C
Vg     1  0        PULSE(-10  10  0  1e-6  1e-6  .01  .02)
R1     1  2        1e3
C1     2  0        1e-6
.TRAN    10e-6    .06
.FOUR    50     v(1)     v(2)
.PROBE
.END
```

Note that in the .TRAN statement above, TSTOP = 0.06 seconds. The choice of a value for TSTOP is based on an analysis of the time constant for this *RC* circuit, which for $C = 1\ \mu\text{F}$ is 1 ms and for $C = 10\ \mu\text{F}$ is 10 ms. Thus, TSTOP is large enough to allow a 6 time constant decay of the natural response for the larger value of *C*, insuring that the Fourier series coefficients will be computed for the steady-state waveform.

The Fourier coefficients for the first nine harmonics for both the input waveform and the output capacitor voltage are:

FOURIER COMPONENTS OF TRANSIENT RESPONSE V(1)

DC COMPONENT = -4.001991E-03

HARMONIC NO	FREQUENCY (HZ)	FOURIER COMPONENT	NORMALIZED COMPONENT	PHASE (DEG)	NORMALIZED PHASE (DEG)
1	5.000E+01	1.273E+01	1.000E+00	-3.600E-02	0.000E+00
2	1.000E+02	8.004E-03	6.286E-04	-9.011E+01	-9.008E+01
3	1.500E+02	4.244E+00	3.333E-01	-1.080E-01	-7.200E-02
4	2.000E+02	8.004E-03	6.286E-04	-9.023E+01	-9.019E+01
5	2.500E+02	2.546E+00	2.000E-01	-1.800E-01	-1.440E-01
6	3.000E+02	8.004E-03	6.286E-04	-9.034E+01	-9.030E+01
7	3.500E+02	1.819E+00	1.429E-01	-2.520E-01	-2.160E-01
8	4.000E+02	8.004E-03	6.287E-04	-9.045E+01	-9.041E+01
9	4.500E+02	1.415E+00	1.111E-01	-3.241E-01	-2.881E-01

```
FOURIER COMPONENTS OF TRANSIENT RESPONSE V(2)


 DC COMPONENT =    9.914491E-04
```

HARMONIC NO	FREQUENCY (HZ)	FOURIER COMPONENT	NORMALIZED COMPONENT	PHASE (DEG)	NORMALIZED PHASE (DEG)
1	5.000E+01	1.217E+01	1.000E+00	-1.755E+01	0.000E+00
2	1.000E+02	1.685E-03	1.384E-04	5.694E+01	7.449E+01
3	1.500E+02	3.083E+00	2.533E-01	-4.425E+01	-2.670E+01
4	2.000E+02	1.223E-03	1.004E-04	3.663E+01	5.418E+01
5	2.500E+02	1.348E+00	1.107E-01	-5.838E+01	-4.083E+01
6	3.000E+02	9.147E-04	7.513E-05	2.590E+01	4.345E+01
7	3.500E+02	7.458E-01	6.126E-02	-6.656E+01	-4.901E+01
8	4.000E+02	7.195E-04	5.910E-05	1.847E+01	3.602E+01
9	4.500E+02	4.581E-01	3.763E-02	-7.310E+01	-5.555E+01

Note that this PSpice analysis confirms our previous hand calculations for the first three odd harmonics. Note further that the even harmonics are essentially zero-valued; the nonzero values that are printed arise because of the imprecision inherent in digital computer calculations.

Now examine the graph of the time response for both the input pulsed waveform and the output capacitor voltage produced by PROBE shown in Fig. 62. Because of the small capacitor value, the output voltage basically is able to follow the input voltage. Finding that the Fourier coefficients of the input and output voltages are so similar is not surprising. Figure 63, produced

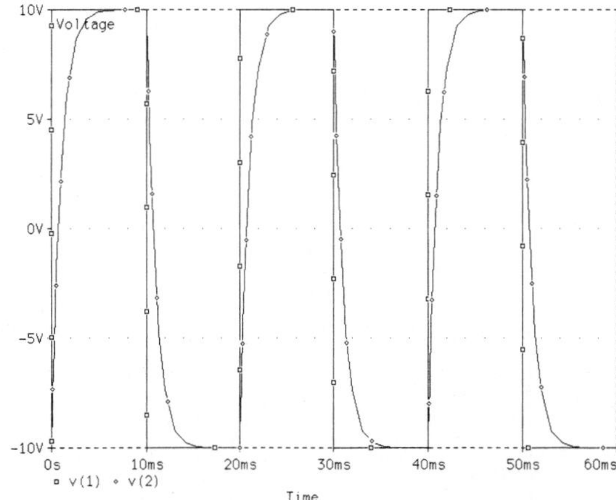

FIGURE 62 PROBE time response plot for the *RC* circuit with small *C*.

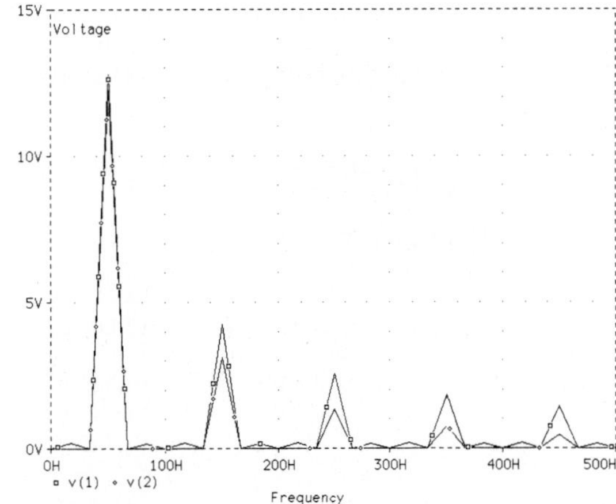

FIGURE 63 PROBE Fourier transform of the waveforms shown in Figure 62.

from Fig. 62 in PROBE by selecting the **Fourier** option from the **X axis** submenu, graphically confirms this similarity.

The PSpice source file for the circuit shown in Fig. 61 with the large capacitor value and the pulsed input shown in Fig. 60 is:

```
Fourier Analysis of RC Circuit with Pulse Input -- Large C
Vg     1  0      PULSE(-10  10  0  1e-6  1e-6  .01  .02)
R1     1  2      1e3
C1     2  0      10e-6
.TRAN    10e-6    .06
.FOUR    50     v(1)     v(2)
.PROBE
.END
```

The resulting Fourier series coefficients for the first nine harmonics of the capacitor voltage, confirming the hand calculations presented, are:

```
FOURIER COMPONENTS OF TRANSIENT RESPONSE V(1)

 DC COMPONENT =  -4.001991E-03

 HARMONIC   FREQUENCY    FOURIER     NORMALIZED     PHASE        NORMALIZED
    NO         (HZ)     COMPONENT    COMPONENT      (DEG)       PHASE (DEG)

     1      5.000E+01   1.273E+01    1.000E+00    -3.601E-02    0.000E+00
     2      1.000E+02   8.004E-03    6.286E-04    -9.011E+01   -9.008E+01
     3      1.500E+02   4.244E+00    3.333E-01    -1.080E-01   -7.202E-02
     4      2.000E+02   8.004E-03    6.286E-04    -9.023E+01   -9.019E+01
     5      2.500E+02   2.546E+00    2.000E-01    -1.801E-01   -1.440E-01
     6      3.000E+02   8.004E-03    6.286E-04    -9.034E+01   -9.030E+01
     7      3.500E+02   1.819E+00    1.429E-01    -2.521E-01   -2.161E-01
     8      4.000E+02   8.004E-03    6.287E-04    -9.045E+01   -9.041E+01
     9      4.500E+02   1.415E+00    1.111E-01    -3.241E-01   -2.881E-01

FOURIER COMPONENTS OF TRANSIENT RESPONSE V(2)

 DC COMPONENT =  -4.143341E-02

 HARMONIC   FREQUENCY    FOURIER     NORMALIZED     PHASE        NORMALIZED
    NO         (HZ)     COMPONENT    COMPONENT      (DEG)       PHASE (DEG)

     1      5.000E+01   3.863E+00    1.000E+00    -7.293E+01    0.000E+00
     2      1.000E+02   1.304E-02    3.374E-03    -1.708E+02   -9.789E+01
     3      1.500E+02   4.460E-01    1.154E-01    -8.556E+01   -1.263E+01
     4      2.000E+02   6.573E-03    1.701E-03    -1.752E+02   -1.022E+02
     5      2.500E+02   1.606E-01    4.156E-02    -8.857E+01   -1.564E+01
     6      3.000E+02   4.386E-03    1.135E-03    -1.765E+02   -1.036E+02
     7      3.500E+02   8.243E-02    2.134E-02    -8.991E+01   -1.698E+01
     8      4.000E+02   3.289E-03    8.512E-04    -1.770E+02   -1.041E+02
     9      4.500E+02   5.077E-02    1.314E-02    -9.101E+01   -1.808E+01
```

PROBE plots of the time response for the input and output voltages, as shown in Fig. 64, illustrate that the large capacitor value prevents the output voltage from following the input voltage. Because the voltage waveforms are so different in the time domain, you should expect very different Fourier coefficients. You saw this difference in both the hand calculations and the PSpice analysis. The PROBE generated plot of the Fourier transformed voltages shown in Fig. 65 further confirms the different spectral content in the two waveforms.

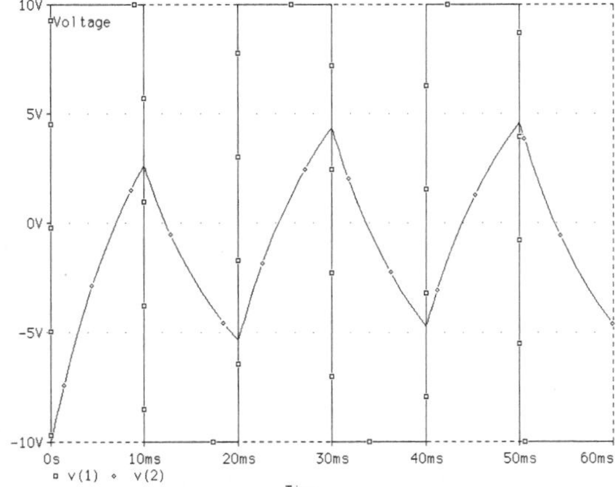

FIGURE 64 PROBE time response plot for the *RC* circuit with large *C*.

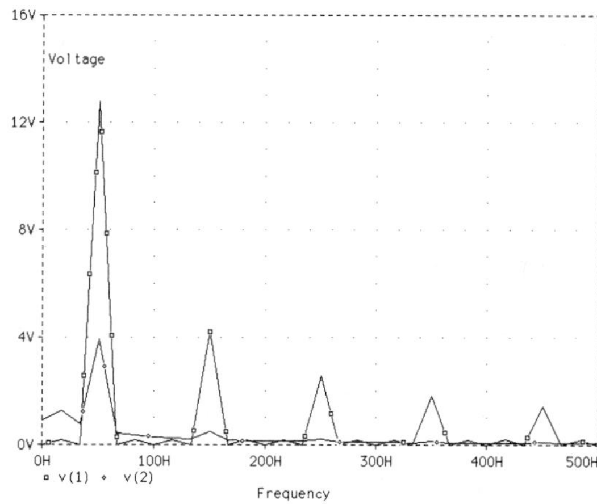

FIGURE 65 PROBE Fourier transform of the waveforms shown in Figure 64.

CHAPTER 18

SUMMARY

PSpice is one of numerous software packages designed to simulate an electric circuit or network. One of the helpful features of these simulation programs is that they allow the user to define the circuit in terms of a topological, as opposed to a mathematical, description. Although the primary purpose of PSpice is to facilitate the analysis of circuits containing nonlinear electronic components, we have introduced it to you as a tool for analyzing linear, lumped-parameter circuits. There are several reasons for this early introduction. First, it acquaints you with a powerful computational tool. Second, you can test the computer-generated values by direct computation, thus gaining confidence in your ability to formulate correct source files. Third, the computer makes it possible for you to attack circuit problems that are computationally very complex and lengthy if attempted without the aid of a computer. These complex problems often add significantly to your understanding of circuit behavior. Finally, simulation software is part of modern engineering practice. The introduction of such software at an elementary level is a first step in preparing you for what lies ahead.

APPENDIX A

GUIDE TO PSPICE CONTROL SHELL AND PROBE MENUS

The software packages PSpice and PROBE are stand-alone executable programs. Usually, you can execute PSpice from the operating system prompt by typing `pspice`. PSpice will then ask you to input the name of the source file you want to simulate. The file name usually has the extension `.CIR`. When PSpice has completed the simulation, it writes the results to a file having the same file name as the source file and the extension `.OUT`. This output file contains a listing of the source code and output from the analysis, including data requested in the `.PRINT` commands and plots generated by the `.PLOT` commands. You may examine this output file on your terminal or by printing it on a line printer or other hard-copy device.

If you have included a `.PROBE` command in your source file, PSpice will also create a file called PROBE.DAT. You may run PROBE to examine your results graphically by typing `probe` at the operating system prompt. PROBE then automatically reads the file named PROBE.DAT produced by PSpice analysis.

MicroSim has bundled PSpice and PROBE together, along with an editor, an on-line help system, an on-line manual, a calculator, and several other helpful tools within the PSpice Control Shell. The Control Shell is menu-based, and may make PSpice and PROBE a bit simpler for you to use. In this appendix we present a brief introduction to the menus that you encounter in the Control Shell and to the menus that you use with PROBE.

The PSpice Control Shell Menus

Normally you enter the PSpice Control Shell[†] by typing `ps` at the operating system prompt, which brings up the start-up screen for the Control Shell. At the bottom of the start-up screen two lines are displayed. The first contains the file that you are currently working with; if there is no current file, the space for the file name will be empty. The second contains the definitions for the function keys, which include F1, F2, F3, F4, F5, F6, and ESC. To access these functions, simply press the desired function key on your keyboard. The following functions are available.

F1—Help Pressing the F1 function key opens a window on the screen containing context-sensitive help. That is, if you are using the calculator and you press F1, a help screen will appear and describe the use of the calculator. Similarly, if you are displaying the submenu for the **Files** option in the Main menu, the help screen describes the submenu selections. You remove the help screen by pressing the ESC key.

F2—Move Pressing the F2 function key allows you to specify a new position for an overlaid window, such as the help screen. Moving an overlaid window sometimes is necessary for you to view the text on the main window.

F3—Manual Pressing the F3 function key activates the on-line manual. This manual describes the syntax for specifying all the components of a PSpice source file. It includes several examples of correct syntax for each PSpice statement. To use this manual outside the editor or the browser, select manual topics from several successive windows. If you are in the editor or the browser, the manual becomes context sensitive. That is, a choice within each window is highlighted to correspond to the type of statement on the current editor or browser line. This aid makes correcting errors in your source file easy, without your having to refer to any PSpice manuals or texts.

F4—Choices Pressing the F4 function key displays the Choices window, which contains a set of responses appropriate to the current context. For example, when the prompt appears for the input of the current PSpice source file, pressing F4 causes all files on the current directory with the extension `.CIR` to be displayed in the Choices window. You may select from among them

---

[*] The following references the PSpice Control Shell, version 4.5 and the Design Center 3, version 5.1, both from MicroSim.

by using the keyboard arrows and pressing enter or by using the mouse.

F5—Calculator Pressing the F5 function key displays the calculator window and an on-line simulation of a hand-held calculator. Pressing the F1 key from the calculator window will tell you how to use the calculator.

F6—Errors The F6 function key invokes the Errors window, which displays any errors encountered when PSpice reads in the current file. Such errors are usually the result of incorrect syntax. You can identify them by displaying the file in the editor and correct them with the help of the on-line manual.

ESC—Cancel Pressing this key causes the current window to be closed and the previous window to be restored.

The Main menu items appear at the top of the start-up screen. They support creation of a PSpice source file and permit you to execute both PSpice and PROBE. The selections include **Files**, **Circuit**, **StmEd**, **Analysis**, **Display**, **Probe**, and **Quit**. Although these Main menu selections are always displayed in the start-up screen, only those items that are highlighted or boldfaced may be selected. To select a menu item, you may type the first letter of the item you want, or you may move the cursor to the menu item you want by using the arrow keys on the keyboard, and hit the Enter key. If you have installed a mouse, you may use the mouse to move to the entry you want, then hit either the left or right mouse key.

FILES

Selecting **Files** opens a pull-down menu with selections that permit you to create and edit a PSpice source file, browse through a PSpice output file, and specify the peripherals available on your computer. The **Files** menu contains the following items.

Edit This option runs the editor built into the PSpice Control Shell on the current file. You may use the arrow keys on the keyboard to position the cursor within the editor window. You may also use several keyboard control sequences to position within the window. Pressing the F1 key while you are in the editor will help you to learn the simple editing commands. Exit the editor with the ESC key.

Browse Output This option opens a window containing the output file for the current file. The output file has the same file name as the current file, but has the extension .OUT. The output file contains the results from a PSpice simulation. You may use the arrow keys and the editing control sequences to move within this output file. The output file is read-only; you are not permitted to make any changes in this file.

Current File This option opens a window which prompts you to input the file name for the current PSpice source file. Press F4 to get a list of all files with the extension `.CIR` on the current directory.

Save File This option allows you to save the contents of the current file to a file name that you input. This procedure is useful if you have made changes to the current file and want to save both the changed and the original files.

X-External Editor This option executes an external editor on the current file. It allows you to use an editor of your choice, rather than the one included with the PSpice Control Shell.

R-External Browser This option executes an external editor on the current output file. You may read the file but not change it in any way. Again, this procedure permits you to use a browser of your choice, rather than the one included with the PSpice Control Shell.

Display/Prn Setup This option opens a window that permits you to specify the display type, the printer port and the printer type for your system. For each entry, pressing the F4 key permits you to choose from a list of valid options for your system.

CIRCUIT

The Main menu **Circuit** selection permits you to make changes to your source file without returning to the editor. You make changes by specifying the type of PSpice statement that you want to modify (Device, Model). Then a window opens to display all the lines in the current file that correspond to the selected statement type. You may choose the specific statement to change by highlighting it with the arrow keys. Any changes that you make to particular PSpice statements are automatically incorporated in the current file. Remember to use the ESC key to exit from each window level.

STMED

The **StmEd** (Stimulus Editor) allows you to specify the form of the independent voltage and current sources in the current file. Only minimal support of the Stimulus Editor is provided with the Student Version of PSpice, so we do not discuss this feature further. If the Stimulus Editor is available on your system, refer to MicroSim's PSpice documentation for a full description of this feature.

ANALYSIS

The **Analysis** pull-down menu allows you to specify the type of analysis PSpice will perform on your source file. You already are familiar with how to specify analysis within the source file; for example, to perform dc analysis, you need a .DC statement in your source file, whereas to perform transient response analysis you need a .TRAN statement. The **Analysis** menu item permits you to specify or change the analysis type without having to enter the editor. Selecting an analysis type automatically adds the appropriate PSpice statement to the current file. The submenu contains the following selections.

Run PSpice This selection runs PSpice on the current source file.

AC & Noise This selection permits you to specify ac analysis and its parameters. A window allows you to enter the values for the .AC statement, which are then written into the current file.

DC Sweep This selection permits you to specify dc analysis and its parameters. You enter parameter values into a window at the prompts. The result is addition of a .DC statement to the current file.

Transient This selection permits you to specify transient analysis or Fourier series analysis and their parameters. Enter parameter values into a window and use them in formulating the .TRAN statement or the .FOUR statement, which is added to the current file.

Parametric This selection permits you to specify the format for a .STEP command, which is added to the current file.

Temperature This selection adds a .TEMP statement to the current file and is beyond the scope of this manual.

Monte Carlo This selection governs Monte Carlo simulation of the current source file and is beyond the scope of this manual.

Options This selection allows you to change or add options and option values to the .OPTIONS statement in the current file. See Appendix B for a definition of this statement.

DISPLAY

The **Display** option allows you to specify output variables to be included in a .PRINT statement. That statement is then added to the current file.

PROBE

The **Probe** option allows you to run PROBE, to add a .PROBE statement to the current file, and to control the contents of the PROBE.DAT file that PROBE uses. The submenu contains the following items.

Run Probe This selection executes PROBE, the graphics post-processor software used to display the results of PSpice simulation.

Auto-run This selection allows you to set a flag that causes PROBE to execute automatically after PSpice analysis. The default value for the flag is YES.

None/Some/All This selection allows you to specify the circuit variables to be included in the PROBE.DAT file used by PROBE. The default is ALL.

Command File This selection allows you to specify the name of a PROBE command file and is beyond the scope of this manual.

Log to File This selection causes PROBE to create a log file. The default is to disable creation of this log file.

Format This selection allows you to specify whether the intermediate file maintained by PROBE is in binary or ASCII format.

QUIT

The **Quit** menu selection allows you to exit the PSpice Control Shell or to execute a single operating system command and return to the shell. Your choice is made from a pull-down menu.

USING THE CONTROL SHELL MENUS

Although the description of the Control Shell menu system might seem a bit imposing, you should play with it in order to become familiar with its capabilities. The on-line Help windows and on-line PSpice manual can assist you greatly in creating a PSpice source file and executing PSpice and PROBE. In just a short time you will be able to master the Control Shell and will perhaps wonder how you could have ever used PSpice without its support. To get you started, we present a sample session.

1. Type `ps` from the operating system to enter the PSpice Control Shell.

2. Select **Files** from the start-up screen menu.
3. Select **Current File** from the Files submenu and type in a file name for your source file at the window prompt. Be sure to include the extension `.CIR` with your file name. Hit the enter (return) key to make this file the current file and to return you to the start-up screen.
4. Select **Files** from the start-up screen menu again.
5. Select **Edit** from the **Files** submenu. If you have specified an existing file name for the current file, the file will appear on the screen. If you are creating a new source file, the screen will be empty.
6. Type in your PSpice source file if this is a new file name, or modify the current file if the file name already exists. Remember to include the `.PROBE` statement if you want to run PROBE to display your results. When you have finished, hit ESC to return to the start-up screen.
7. Select **Analysis** from the start-up menu.
8. Select **Run PSpice** from the **Analysis** submenu. PSpice then executes and displays screens that indicate its progress in simulating the circuit you specified in the source file. When PSpice has finished, it returns you to the start-up screen.
9. Select **Files** from the start-up menu.
10. Select **Browse Output** from the **Files** submenu and read through the output file produced by PSpice. If you want to print the output file, hit ESC twice to exit the browser and the **Files** submenu. Then select **Quit** from the start-up menu and **DOS Command** from the **Quit** submenu and input `type` *filename*`.OUT` at the window prompt, where *filename* is the name of the current file. When printing begins you return to the Control Shell by hitting the enter (return) key.
11. Select **Probe** from the start-up menu.
12. Select **Run Probe** from the **Probe** submenu and use the **Probe** menu items to create your plots. The **Probe** menu item **Exit** will return you to the PSpice Control Shell.
13. To exit from the PSpice Control Shell, select **Quit** from the start-up menu and **Exit to DOS** from the **Quit** submenu. You will then see the operating system prompt.

The PROBE Menu System

PROBE is a graphics postprocessor software package that reads a special file written by PSpice (if you have included the .PROBE statement in your PSpice source file) and displays PSpice-computed voltages and currents in a graphic format. You can execute PROBE from the PSpice Control Shell or from the operating system prompt. PROBE is menu-driven; when you invoke PROBE, a blank plot appears on the screen, with the Main menu options at the bottom. You select an option from the Main menu by typing the first letter of the option name, using the keypad arrows to highlight the desired option, and pressing the enter key,—or by using a mouse if one has been installed. Each option from the Main menu produces a submenu from which you may further specify your choice. In the following sections we describe the Main menu and submenu options that are useful in displaying simulation results for electric circuits constructed from the tools described in the Nilsson text. Menu options that control the display of digital data, Monte Carlo analysis, and the like are beyond the scope of this manual. MicroSim's PSpice Reference Manual contains detailed descriptions of PROBE's capabilities.

The Start-Up Menu

The Start-Up menu allows you to select the type of analysis that generated the data you want PROBE to plot. If only one type of analysis was specified in the PSpice source file, you will not see the start-up menu. If you have specified two or more types of analysis in the source file, you must choose which one to look at. The following start-up menu options are available.

Exit program This option terminates PROBE and returns you to the Control Shell or to the operating system, depending on where you invoked PROBE.

Dc sweep This option allows you to plot the results from an analysis specified with the .DC command. It invokes the Analog Plot menu.

Ac sweep This option allows you to plot the results from an analysis specified with the .AC command. It also invokes the Analog Plot menu.

Transient analysis This option allows you to plot the results from an analysis specified with the .TRAN command. It also invokes the Analog Plot menu.

THE ANALOG PLOT MENU

The Analog Plot menu is the main menu in PROBE and is displayed at the bottom of the plot screen when it first appears. Each menu item invokes a submenu from which you can select further. The menu and submenus contain the following items.

Exit This selection exits the Analog Plot menu and returns you to the Start-Up menu.

Add trace This selection adds one or more traces to the current plot. If more than one plot is displayed, the trace is added to the plot currently selected (identified by the symbol SEL>> next to the plot). The traces can be any output variables resulting from PSpice simulation or an arithmetic expression involving the output variables. Pressing the F4 key will display a list of valid variables from which to select. Refer to PROBE documentation for examples of the use of arithmetic expressions.

Remove trace This selection removes one or more traces from the selected plot. A displayed submenu allows you to select **All** or specific individual traces for removal.

X axis This selection invokes the *x*-axis submenu. The submenu options are similar to those in the *y*-axis submenu. We describe the axis submenu items after the Y **axis** entry.

Y axis This selection invokes the *y*-axis submenu. Both the *x*-axis and *y*-axis submenus contain the following selections.

- **Exit** This selection returns you to the Analog Plot menu.
- **Log** This selection specifies a logarithmic axis. For ac analysis, the *x* axis is already logarithmic.
- **Linear** This selection specifies a linear axis. Except for the *x* axis in ac analysis, all axes start out linear.
- **Auto range** This selection asks PROBE to choose a convenient range of values to be displayed on the axis. All axes start out autoranging and this submenu item does not appear at the start.
- **Set range** This selection allows you to enter a minimum and maximum value for the range of values displayed on the axis.
- **Restrict data** This selection appears on the *x*-axis submenu only. It is used to restrict the range of *x*-axis data used in the computation of the Fourier transform coefficients.

- **Unrestrict data** This selection appears only if the **Restrict data** item has been previously selected from the *x*-axis submenu. It is used to undo the effect of the **Restrict data** input.
- **X variable** This selection sets the variable for the *x* axis; it appears in the submenu of the **X axis** menu item only. The *x*-axis variable default depends on the analysis type: voltage or current for a dc sweep, frequency for ac analysis, and time for transient analysis. You use this submenu item to override the default variable for the *x* axis. A prompt will ask you to enter the variable to be displayed on the *x* axis. This option allows you to plot any variable versus any other variable for a given analysis.
- **Fourier** This selection invokes the Fourier transform mode of PROBE. All traces currently displayed are transformed. If the *x*-axis variable is time, a forward Fourier transform converts the *x*-axis variable to frequency; if the *x*-axis variable is frequency, an inverse Fourier transform converts the *x*-axis variable to time.
- **Quit fourier** This selection reverses the transformation performed by the **Fourier** selection and returns the *x*-axis variable to its original state.

Plot control This selection invokes the plot control submenu. This submenu contains many items, including the following, which are most often used.

- **Exit** This selection returns you to the Analog Plot menu.
- **Add plot** This selection adds a plot to the top of the screen and selects the new plot. A maximum of 20 plots may be displayed. Each plot has the same *x* axis but can have a different *y* axis.
- **Remove plot** This selection removes the selected plot from the display.
- **Select plot** This selection allows you to change the selected plot using the arrow keys.

Display Control This selection invokes the display control submenu and provides a method for modifying display attributes, which is beyond the scope of this manual.

Macros This selection invokes the macro control submenu, which is beyond the scope of this manual.

Hard Copy This selection generates a hard copy of the current display on the plotting device, which usually is a printer with graphics capabilities.

Cursor This selection gives you control over two screen cursors that you can use to read values from the selected plot. You activate the first cursor by using the left and right arrow keys. You activate the second cursor by holding down the SHIFT key while using the left and right arrow keys. A display window in the lower right corner contains the current values at the location of the two cursors. A cursor submenu provides specific cursor placement commands, such as **Peak**, **Trough**, **Min** and **Max**, which are beyond the scope of this manual.

Zoom This selection allows you to identify a section of the plot and zoom in or out. The effect is the same as changing the variable range on the *x* and *y* axes. The **zoom** submenu includes the following items.

- **Specify region** After you have selected this submenu item, press the spacebar to identify one corner of the zoom region. Then use the arrow keys or the mouse to move the cursor to the opposite corner and press the spacebar a second time. This creates the zoom box and causes the display to be redrawn within that box.
- **X zoom in** This item causes the *x* axis to be zoomed in by a factor of 2 on the currently selected plot.
- **Y zoom in** This item causes the *y* axis to be zoomed in by a factor of 2 on the currently selected plot.
- **Zoom out** This item causes the display to be zoomed out by a factor of 2 in both the *x* and *y* directions.
- **Pan** This item pans various portions of the screen by 10% in the direction specified by the arrow key being pressed.
- **Auto range** This item undoes all previous effects of zooming and panning.

Label This selection enables you to label your plot with both symbols and text by selecting various Label submenu items. The items you will use most often are:

- **Exit** This item exits from the Label submenu back to the Analog Plot menu.
- **Text** This item allows you to enter text at a prompt and use the arrow keys to place that text anywhere on the plot.

- **Line** This item allows you to create a single straight line anywhere on the plot by specifying the endpoints of the line with the arrow keys.
- **Delete one** This item allows you to delete any single element previously entered on the plot from the Label submenu. You use the arrow keys to position the cursor over the item to be deleted.
- **delete All** This item allows you to delete all elements entered on the plot from the Label submenu.
- **Move** This item allows you to identify any previously entered plot element and move it to a new location. The arrow keys identify the item and also identify its new location.

USING THE PROBE MENUS

As was the case with the PSpice Control Shell menu system, the PROBE menus may seem complicated and confusing at first. But the only menu option you actually need at first is **Add Trace**. You should play with the PROBE menus and discover how the menu items affect the displayed trace. In a short time you will be using PROBE in sophisticated ways to provide added insight into the functions of the circuit you are studying.

APPENDIX B

SPECIFYING PSPICE SIMULATION PARAMETERS

You can use two PSpice statements to control parameters for PSpice simulation and the output file that PSpice produces. The first of these statements is .WIDTH, which controls the number of characters printed in a row of the output file and has the statement syntax:

```
.WIDTH     OUT = value
```

where value is the number of columns in a row and must be 80 (the default) or 132.

The second statement is somewhat more complicated. The .OPTIONS statement allows you to change some of the default values for parameters used in circuit simulations. The format for this statement is:

```
.OPTIONS   OPT1   <OPT2  ...>   <OPTn = value  ...>
```

Options come in two varieties: those that do not have values associated with them and those that do have values. Options without values usually are called flags. The following flags are relevant to the use of PSpice described in this manual.

ACCT Includes accounting information at the end of the output file.

EXPAND Creates a list of devices from expansion of subcircuits.

LIBRARY Includes all lines used in the source file from PSpice libraries.

LIST Creates a summary of circuit elements.

NOBIAS Bias values for node voltages are not printed in the output file.

NODE Includes a node table to summarize circuit connections.

NOECHO Source file listing is not included in the output file.

NOMOD Model parameters are not listed in the output file.

NOPAGE Form feeds and headers are not included in the output file.

OPTS Includes a list of all flags and option values in the output file.

WIDTH Sets the width of the output file to 132 columns.

The following options with values are described with their default values and units. This partial list identifies the options that are relevant to the PSpice description included in this manual. See the MicroSim PSpice documentation for a complete list of all options.

ABSTOL = (1 pA) Sets the error tolerance on circuit currents.

CPTIME = (1E6 sec) Sets the CPU time allotted for the simulation.

GMIN = (1E-12 ohm^{-1}) Sets the minimum conductance in a circuit branch.

ITL1 = (40) Sets the iteration limit for dc and bias-point computation when the source file provides no starting point.

ITL2 = (20) Sets the iteration limit for dc and bias-point computation when the source file provides a starting point.

ITL4 = (10) Sets the iteration limit for the computation of any point in a transient analysis.

ITL5 = (5000) Sets the total number of iterations allowed for all points in a transient analysis. Setting ITL5 = 0 removes the limit on iterations.

LIMPTS = (∞) Sets the maximum number of points to be printed in the output or on a plot.

NUMDGT = (4) Sets the number of digits used for printing numbers in the output file.

PIVREL = (1E-3) Sets the absolute minimum value for a matrix entry to be accepted as a pivot.

PIVTOL = (1E-13) Sets the relative ratio between the largest column entry and an acceptable pivot value.

RELTOL = (.001) Sets the relative error tolerance for voltages and currents.

TNOM = (27° C) Sets the nominal temperature.

VNTOL = (1 μV) Sets the error tolerance on circuit voltages.

APPENDIX C

DIFFERENCES BETWEEN SPICE AND PSPICE

The differences between SPICE (that is, the software package SPICE2G.6 from the University of California at Berkeley) and PSpice fall into two categories. The first category contains the differences that you will encounter in attempting to use PSpice to simulate a circuit described by a source file originally prepared for SPICE. The second category contains the differences you will encounter in going in the other direction—that is, in attempting to use SPICE to simulate a circuit described by a source file originally prepared for PSpice. In the discussion that follows, we identify only those differences relevant to using SPICE and PSpice for simulating circuits from material in the Nilsson text. Other differences exist with regard to both analog and digital electronic circuits, but we do not cover them here. Refer to the MicroSim PSpice documentation appendices for the complete list of differences.

With the following exceptions, any source file developed for SPICE can be successfully read and simulated by PSpice.

1. PSpice does not support the following option fields in the `.OPTION` statement: `LIMTIM`, `LVLCOD`, `METHOD`, `MAXORD`, `LVLTIM`, and `ITL3`.

2. PSpice does not support the IN= option for the `.WIDTH` statement.

3. Voltage coefficients for capacitors and current coefficients for inductors must be incorporated into the `.MODEL` statement rather than into the statement defining the device.

In general, SPICE can read and simulate a source file developed for PSpice. However, some features available in PSpice are not supported by SPICE. Any source file that uses any of the following features would have to be modified before SPICE simulation.

1. SPICE does not support voltage and current controlled switches.
2. SPICE does not support models for resistors, inductors, and capacitors, such as those used in conjunction with the `.STEP` command.
3. SPICE does not allow logarithmic sweeps of dc voltages or currents.
4. SPICE does not support sweeps of parameters in a `.MODEL` statement.
5. SPICE does not support the `.LIB` statement.
6. All characters in a SPICE source file must be uppercase.

APPENDIX D

QUICK REFERENCE TO PSPICE

DATA STATEMENTS

In the syntax for data statements, xxx represents any alphanumeric string used to identify uniquely a circuit component, n+ represents the positive circuit node and n− represents the negative voltage node. Positive voltage drop is from n+ to n−, and positive current flows from n+ to n−. Any elements in the syntax specification that are optional are enclosed by brackets ([ ]).

INDEPENDENT DC VOLTAGE AND CURRENT SOURCES

SYNTAX (VOLTAGE) Vxxx n+ n− DC value

SYNTAX (CURRENT) Ixxx n+ n− DC value

where

value is the dc voltage in volts for the voltage source and the dc current in amperes for the current source.

DESCRIPTION Provides a constant source of voltage or current to the circuit.

EXAMPLES

```
Vin 1 2 DC 3.5

Isource 3 0 DC 0.25
```

Independent AC Voltage and Current Sources

Syntax (voltage) Vxxx n+ n− AC mag [phase]

Syntax (current) Ixxx n+ n− AC mag [phase]

where

mag is the magnitude of the ac waveform in volts for the voltage source and in amperes for the current source;

phase is the phase angle of the ac waveform in degrees and has a default value of 0.

Description Supplies a sinusoidal voltage or current at a fixed frequency to the circuit.

Examples

```
Vab 3 2 AC 10 90

I12 4 6 AC 0.5
```

Independent Transient Voltage and Current Sources

Syntax (voltage) Vxxx n+ n− trans_type

Syntax (voltage) Ixxx n+ n− trans_type

where trans_type is one of the following transient waveform types.

- *Exponential* EXP (start peak [delay1] [rise] [delay2] [fall]) with

  start The initial value of voltage in volts or current in amperes;

  peak The maximum value of voltage in volts or current in amperes;

  delay1 The delay in seconds prior to a change in voltage or current from the initial value, which is 0 seconds by default;

  rise The time constant in seconds of the exponential decay from start to peak, which has a default value of tstep seconds (see . TRAN);

  delay2 The delay in seconds prior to introducing a second exponential decay, from the maximum value toward the initial value, which has a default value of delay1 + tstep seconds (see . TRAN);

fall The time constant in seconds of the exponential decay from peak to final, which has a default value of tstep seconds (see . TRAN).

- *Pulsed* PULSE(min max [delay] [rise] [fall] [width] [period])

with

min The minimum value of the waveform in volts for the voltage source and in amperes for the current source;

max The maximum value of the waveform in volts for the voltage source and in amperes for the current source;

delay The time in seconds prior to the onset of the pulse train, which has a default of 0 seconds;

rise The time in seconds for the waveform to transition from min to max, which has a default value of tstep seconds (see . TRAN);

fall The time in seconds for the waveform to transition from max to min, which has a default value of tstep seconds (see . TRAN);

width The time in seconds that the waveform remains at the max value, which has a default value of tstop seconds (see . TRAN);

period The time in seconds that separates the pulses in the pulse train, which has a default of tstop seconds (see . TRAN).

- *Piecewise linear* PWL(t1 val1 t2 val2 . . .)

with

t1 val1 A paired time (in seconds) and value (in volts for a voltage source and amperes for a current source) that specifies a corner of the waveform; all pairs of times and values are linearly connected to form the whole waveform.

- *Damped sinusoid* SIN (off peak [freq] [delay] [damp] [phase])

with

off The initial value of the voltage in volts or the current in amperes;

peak The maximum amplitude of the voltage in volts or the current in amperes;

freq The sinusoidal frequency in hertz, which has a default value of 1/tstop hertz (see . TRAN);

delay The time in seconds that the waveform stays at the initial value before the sinusoidal oscillation begins, which has a default value of 0 seconds;

damp The sinusoidal damping factor in seconds^{-1} used to specify the decaying exponential envelope for the sinusoid, which has a default value of 0 seconds^{-1};

phase The initial phase angle of the sinusoidal waveform in degrees, which has a default value of 0 degrees.

DESCRIPTION Supplies a time-varying voltage or current to the circuit whose waveform can be characterized as exponential, pulsed, piecewise linear, or damped sinusoidal.

EXAMPLES

```
Vs 3 1 EXP(2 6 .5 .1 .5 .2)

Iin 4 0 PULSE(-.3 .3 0 .01 .01 1 2)

I5 2 3 PWL(0 .2 1 .6 1.5 .6 3 -.5)

Vg 0 3 SIM(0 2 100 0 0 90)
```

DEPENDENT VOLTAGE CONTROLLED VOLTAGE SOURCE

SYNTAX Exxx n+ n− cn+ cn− gain

where

cn+ is the positive node for the controlling voltage;

cn− is the negative node for the controlling voltage;

gain is the ratio of the source voltage (between n+ and n−) to the controlling voltage (between cn+ and cn−).

DESCRIPTION Provides a voltage source whose value depends on a voltage measured elsewhere in the circuit.

EXAMPLE

```
Eop 1 2 4 0 .5
```

DEPENDENT CURRENT CONTROLLED CURRENT SOURCE

SYNTAX Fxxx n+ n− Vyyy gain

where

Vyyy is the name of the voltage source through which the controlling current flows;

gain is the ratio of the source current (flowing from n+ to n−) to the controlling current (flowing through Vyyy).

DESCRIPTION Provides a current source whose value depends on the magnitude of a current flowing through a voltage source elsewhere in the circuit.

EXAMPLE `Fdep 3 2 Vcontrol 10`

VOLTAGE CONTROLLED CURRENT SOURCE

SYNTAX Gxxx n+ n− cn+ cn− gain

where

cn+ is the positive node for the controlling voltage;

cn− is the negative node for the controlling voltage;

gain is the ratio of the source current (flowing from n+ to n−) to the controlling voltage (defined between cn+ and cn−), in ohms^{-1}.

DESCRIPTION Provides a current source whose value depends on the magnitude of a voltage measured elsewhere in the circuit.

EXAMPLE `Gon 3 6 4 1 0.35`

CURRENT CONTROLLED VOLTAGE SOURCE

SYNTAX Hxxx n+ n− Vyyy gain

where

Vyyy is the name of the voltage source through which the controlling current flows;

gain is the ratio of the source voltage (defined between n+ and n−) to the controlling current (flowing through Vyyy), in ohms.

DESCRIPTION Provides a voltage source whose value depends on the magnitude of current flowing through another voltage source elsewhere in the circuit.

EXAMPLE `Hout 7 2 Vdummy −2.5e−3`

Voltage Controlled Switch

Syntax Sxxx n+ n− cn+ cn− mname

where

cn+ is the positive node of the controlling voltage;

cn− is the negative node of the controlling voltage;

mname is the name of a VSWITCH model defined in a .MODEL statement (see .MODEL).

Description Simulates a switch that is opened or closed, depending on the magnitude of a voltage measured elsewhere in the circuit.

Examples
```
S32  13  12  4  0  smod
.MODEL  smod  VSWITCH(RON = 0.1)
```

Current Controlled Switch

Syntax Wxxx n+ n− Vyyy mname

where

Vyyy is the name of the voltage source through which the controlling current flows;

mname is the name of an ISWITCH model defined in a .MODEL statement (see .MODEL).

Description Simulates a switch that is open or closed, depending on the magnitude of the current flowing through a voltage source located elsewhere in the circuit.

Example
```
Woff  4  8  Vab  imod
.MODEL imod ISWITCH(ROFF = 1e+3, ION = 1e-6)
```

Resistor

Syntax Rxxx n+ n− [mname] value

where

mname is the name of a RES model defined in a .MODEL statement (see .MODEL)—note that mname is optional;

value is the resistance, in ohms.

DESCRIPTION Models a resistor, a circuit element whose voltage and current are linearly dependent.

EXAMPLE `Rfor 3 4 16e+3`

INDUCTOR

SYNTAX `Lxxx n+ n- [mname] value [IC = icval]`

where

mname is the name of an IND model defined in a .MODEL statement (see .MODEL)—note that mname is optional;

value is the inductance, in henries;

icval is the initial value of the current in the inductor, in amperes—note that specifying the initial condition is optional.

DESCRIPTION Models an inductor, a circuit element whose voltage is linearly dependent on the derivative of its current.

EXAMPLE `L44 1 9 3e-3 IC = 1e-2`

CAPACITOR

SYNTAX `Cxxx n+ n- [mname] value [IC = icval]`

where

mname is the name of a CAP model defined in a .MODEL statement (see .MODEL)—note that mname is optional;

value is the capacitance, in farads;

icval is the initial value of the voltage across the capacitor, in volts—note that specifying the initial condition is optional.

DESCRIPTION Models a capacitor, a circuit element whose current is linearly dependent on the derivative of its voltage.

EXAMPLES

```
Ctwo 4 0 cmod 2e-6
.MODEL cmod CAP(C = 1)
```

MUTUAL INDUCTANCE

SYNTAX Kxxx Lyyy Lzzz value

where

Lyyy is the name of the inductor on the primary side of the coil;

Lzzz is the name of the inductor on the secondary side of the coil;

value is the mutual coupling coefficient, k, which has a value such that $0 \leq k < 1$.

DESCRIPTION Models the magnetic coupling between any two inductor coils in a circuit.

EXAMPLE
```
Kab La Lb 0.5
```

SUBCIRCUIT DEFINITION

SYNTAX .SUBCKT name [nodes] .ENDS

where

name is the name of the subcircuit, as referenced by an X statement (see subcircuit call);

nodes is the optional list of nodes used to identify the connections to the subcircuit;

.ENDS signifies the end of the subcircuit definition.

DESCRIPTION Used to provide "subroutine"-type definitions of portions of a circuit. When the subcircuit is referenced in an X statement, the definition between the .SUBCKT statement and the .ENDS statement replaces the X statement in the source file.

EXAMPLE
```
.SUBCKT opamp 1 2 3 4 5
   ⋮
.ENDS
```

SUBCIRCUIT CALL

SYNTAX Xxxx nodes name

where

nodes is the optional list of circuit nodes used to connect the subcircuit into the rest of the circuit; there must be as many nodes in this list as there are in the subcircuit definition (see Subcircuit Definition);

name is the name of the subcircuit as defined in a .SUBCKT statement.

DESCRIPTION Replaces the X statement with the definition of a subcircuit, which permits a subcircuit to be defined once and used many times within a given source file.

EXAMPLE `Xamp 4 2 8 5 1 opamp`

LIBRARY FILE

SYNTAX .LIB [fname]

where

fname is the name of the library file containing .MODEL or .SUBCKT statements referenced in the source file, which by default is the nominal or evaluation library file.

DESCRIPTION Used to reference models or subcircuits.

EXAMPLE `.LIB mylib.lib`

DEVICE MODELS

SYNTAX .MODEL mname mtype [(par = value)]

where

mname is a unique model name, which is also used in the device statement that incorporates this model;

mtype is one of the model types available;

par = value is an optionally specified list of parameters and their assigned values, specific to the model type:

- VSWITCH, which models a voltage-controlled switch and has the following parameters:

  RON, the resistance of the closed switch, whose default value is 1 Ω;
  ROFF, the resistance of the open switch, whose default value is 1e6 Ω;
  VON, the control voltage level necessary to close the switch, whose default value is 1 V;
  VOFF, the control voltage level necessary to open the switch, whose default value is 0 V;

- ISWITCH, which models a current controlled switch and has the following parameters:
  RON, the resistance of the closed switch, whose default value is 1 Ω;
  ROFF, the resistance of the open switch, whose default value is 1e6 Ω;
  ION, the control current necessary to close the switch, whose default value is 1e-3 A;
  IOFF, the control current necessary to open the switch, whose default value is 0 A;
- RES, which models a resistor and has the parameter
  R, the resistance multiplier, whose default value is 1;
- IND, which models an inductor and has the parameter
  L, the inductance multiplier, whose default value is 1;
- CAP, which models a capacitor and has the parameter
  C, the capacitance multiplier, whose default value is 1.

DESCRIPTION Defines standard devices that can be used in a circuit and sets parameter values that characterize the specific device being modeled.

EXAMPLES

```
.MODEL  lmodel  IND(L = 2)

.MODEL  resist  RES
```

CONTROL STATEMENTS

DC ANALYSIS

SYNTAX .DC [type] vname start end incre [nest-sweep]

where

type is the sweep type and must be LIN for a linear sweep from the start value to the end value, OCT for a logarithmic sweep from start value to end value in octaves, DEC for a logarithmic sweep from start value to end value in decades, or LIST if a list of values is to be used; the default sweep type is LIN;

vname is the name of the circuit element whose value is swept, usually an independent voltage or current source;

start is the beginning value of vname;

end is the final value of vname; note that when the sweep type is LIST, the start, end, and incre values are replaced by a list of values that vname will take on;

incre is the step size for the LIN sweep type and is the number of points per octave or decade for the OCT and DEC sweep types;

nest-sweep is an optional additional sweep of a second circuit variable, which follows the same syntax as the sweep for the first variable.

DESCRIPTION Provides a method for varying one or two dc source values and analyzing the circuit for each sweep value.

EXAMPLES

```
.DC Vab 1 10 .5

.DC I1 DEC .1 10 30 V1 DEC 10 100 30
```

TRANSFER FUNCTION

SYNTAX .TF output input

where

output is the output circuit variable, which is a voltage or a current;

input is the source circuit variable, which is a voltage or a current.

DESCRIPTION Computes the gain, input resistance and output resistance between input and output.

EXAMPLE

```
.TF V(6,2) Vsource
```

SENSITIVITY

SYNTAX .SENS vname

Where vname is the circuit variable name (or list of circuit variable names) for which sensitivity analysis will be performed.

DESCRIPTION Computes and prints to the output file a dc sensitivity analysis of vname to the values of other circuit elements, including resistors, independent voltage and current sources, and voltage and current controlled switches.

EXAMPLE

```
.SENS V(2) I(Vin)
```

TRANSIENT ANALYSIS

SYNTAX .TRAN tstep tstop [npval] istep [UIC]

where

tstep is the time interval separating values that are printed or plotted;

tstop is the ending time for the transient analysis;

npval is the time between 0 and the first value printed or plotted, which by default is 0;

istep is the internal time step used for computing values, which by default is tstop/50;

UIC will bypass the calculation of the bias point, which usually precedes the transient analysis, and use the initial conditions specified by IC = in inductor and capacitor data statements instead.

DESCRIPTION Performs a transient analysis of the circuit described in the source file to calculate the values of circuit variables as a function of time.

EXAMPLE .TRAN .01 10 0 .001 UIC

INITIAL CONDITION

SYNTAX .IC Vnode = value

where

Vnode = value is one or a list of pairs consisting of a voltage node Vnode, represented in standard form, and the initial value in volts at that node.

DESCRIPTION Sets the initial conditions for transient analysis by specifying one or more node voltage values.

EXAMPLE .IC V(1) = 10 V(3,4) = −2

AC ANALYSIS

SYNTAX .AC [type] num start end

where

type is the type of sweep, which must be one of the following keywords:

LIN for a linear sweep in frequency, which is the default;

OCT for a logarithmic sweep in frequency by octaves;

DEC for a logarithmic sweep in frequency by decades;

num is the total number of points in the sweep for a linear sweep and the number of points per octave or decade for a logarithmic sweep;

start is the starting frequency, in hertz;

end is the final frequency, in hertz.

DESCRIPTION Computes the frequency response of the circuit described in the source file as the frequency is swept either linearly or logarithmically from an initial value to a final value.

EXAMPLE `.AC LIN 300 10 1000`

STEPPED VALUES

SYNTAX .STEP [type] name start end incre

where

type is one of the following keywords describing the type of parameter variation:

LIN for a linear variation, which is the default;

OCT for a logarithmic variation by octaves;

DEC for a logarithmic variation by decades;

name is the name of the circuit element whose value is to be varied;

start is the starting value for the circuit element, in units appropriate to the type of circuit element;

end is the final value for the circuit element, in units appropriate to the type of circuit element;

incre is the step size for a linear variation and the number of values per octave or decade for the logarithmic variation. Note that when a discrete list of parameter values is desired, type is not used and the name is followed by the keyword LIST and a list of parameter values.

DESCRIPTION Used to step a circuit element's value through a range either linearly or logarithmically or through a discrete list, analyzing the circuit for each value.

EXAMPLES .STEP DEC RES RMOD(R) 10 10e4 3

.STEP V2 LIST 1 4 12 19

FOURIER SERIES ANALYSIS

SYNTAX .FOUR freq vname

where

freq is the fundamental frequency of the circuit;

vname is the variable name or the list of variable names for which Fourier Series coefficients will be computed.

DESCRIPTION Uses the results of a transient analysis to compute the Fourier coefficients for the first nine harmonics. Note the .FOUR statement requires a .TRAN statement to perform the transient analysis.

EXAMPLE .FOUR 10e3 V(3) V(6,2)

OUTPUT STATEMENTS

OPERATING POINT

SYNTAX .OP

DESCRIPTION Outputs information describing the bias point computations for the circuit being simulated.

EXAMPLE .OP

PRINT RESULTS

SYNTAX .PRINT type vname

where

type identifies the type of analysis performed to generate the data being printed and must be one of the keywords DC, AC, or TRAN;

vname is the variable name or the list of variable names for which values are to be printed.

DESCRIPTION Prints the results of circuit analysis in table form to an output file for each program variable specified.

EXAMPLE `.PRINT AC V(3,2) I(Vsource)`

PLOT RESULTS

SYNTAX .PLOT type vname [low] [high]

where

type identifies the type of analysis used to produce the values being plotted and must be one of the keywords DC, AC, or TRAN;

vname is the variable name or the list of variable names for which values are to be plotted;

low and high optionally sets the range of values of the independent variable to be plotted, which defaults to the entire range of values.

DESCRIPTION Generates line-printer plots of circuit variables computed during dc, ac, or transient analysis.

EXAMPLE `.PLOT TRANS I(Vin) V(4)`

PROBE

SYNTAX .PROBE [vname]

Where vname is the optionally specified variable or list of variables whose values from dc, ac, and transient analysis will be stored in the file that PROBE uses to generate plots. If no variable name is included, values of all circuit variables will be stored in the PROBE file.

DESCRIPTION Generates a file of data from dc, ac, and transient analysis used by PROBE to generate high-quality plots.

EXAMPLE `.PROBE V(2) V(3) V(4) I(Rin) I(Rout)`

MISCELLANEOUS STATEMENTS

TITLE LINE

DESCRIPTION Each source file must have a title line as its first line.

EXAMPLE
```
A Circuit to Simulate RLC
*Frequency Response
```

END STATEMENT

SYNTAX .END

DESCRIPTION Each source file must have an .END statement as its final line.

EXAMPLE .END

PRINTING CONTROL

SYNTAX .WIDTH OUT = value

where

value is used to specify the number of columns used in printing the output file, which must be either 80 (the default) or 132.

DESCRIPTION Used to change the number of columns in the output file from the default.

EXAMPLE .WIDTH OUT = 132

SIMULATION OPTIONS

SYNTAX .OPTIONS oname [oname = value]

where

oname is the name or the list of names of option flags to be set;

oname = value is the name and assigned value or the list of names and their values for options requiring values.

DESCRIPTION Used to override the default specifications of many parameters used to control circuit simulation.

EXAMPLE .OPTION NOPAGE NONODE RELTOL = .1

PROBLEMS

1. Use PSpice to find the voltages V(2,0), V(3,0), and V(4,0) in the circuit shown in Fig. P.1.

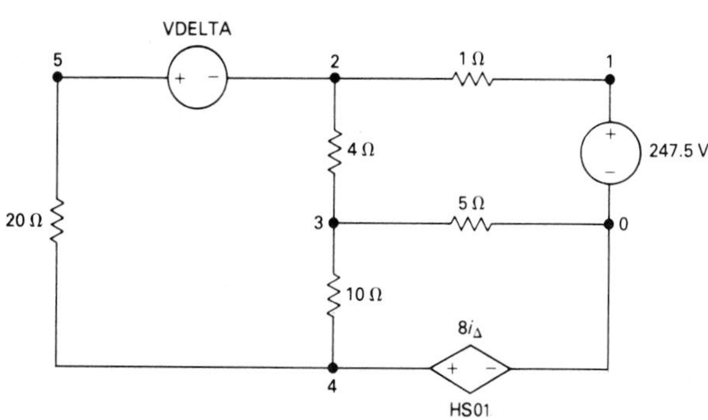

FIGURE P.1

2. a) Use the output from PSpice simulation of the circuit in Problem 1 to calculate the power dissipated in each resistor in Fig. P.1.
   b) Find the total power supplied by the independent voltage source.
   c) Find the total power supplied by the current controlled voltage source.
   d) Verify that the power dissipated equals the power supplied.
   e) What is the total power dissipated value output from the PSpice analysis?
   f) How does the value in (e) compare to the previous power computations?

3. Return to Fig. P.1 and replace the current controlled voltage source with an independent voltage source of 132 V, positive at the reference node. Use PSpice to simulate this circuit and compare the output with that obtained in Problem 1. In particular, compare the total power dissipated values from each simulation.

4. Use PSpice to find the voltages v_a, v_b, and v_c in the circuit shown in Fig. P.4. Use the .DC control statement so that these voltages can be specified directly in a .PRINT statement.

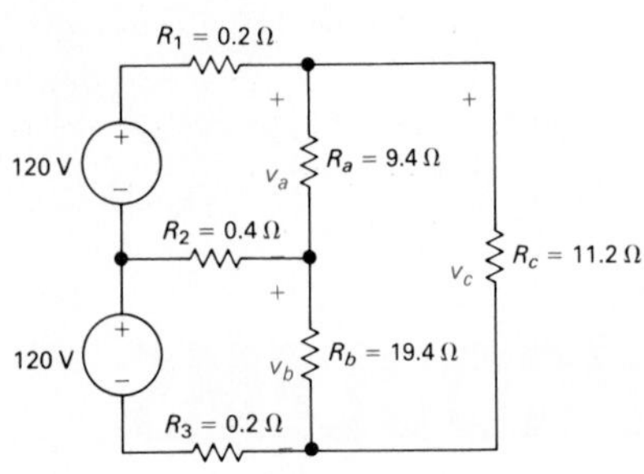

FIGURE P.4

5. Use PSpice to find the Thévenin equivalent with respect to the terminals of the variable resistor R_o in the circuit shown in Fig. P.5.

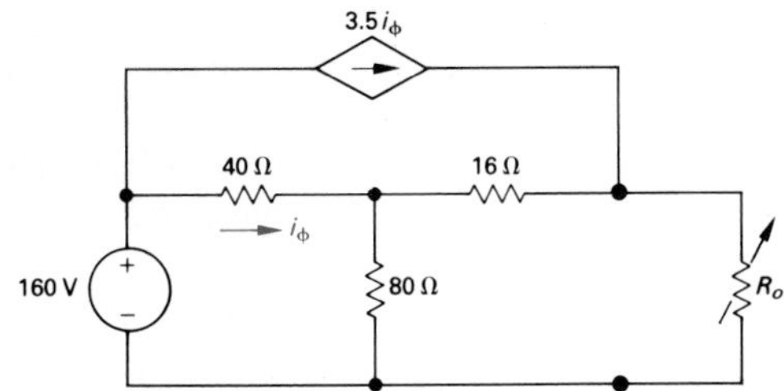

FIGURE P.5

6. a) Label the terminals of the 10-Ω resistor in the circuit shown in Fig. P.6 as a and b. Let a be the terminal at the top of the resistor. Now remove the 10-Ω resistor and use PSpice to find the Thévenin equivalent of the remaining circuit with respect to terminals a, b.

   b) Use the Thévenin equivalent obtained in (a) to find the current in the 10-Ω resistor.

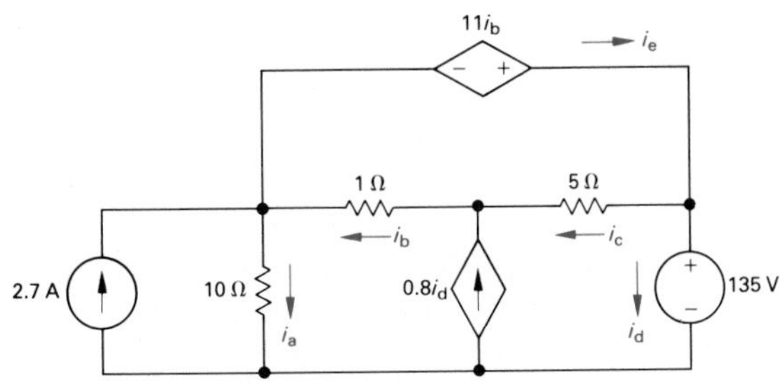

FIGURE P.6

7. Use PSpice to find the Thévenin equivalent with respect to terminals a, b for the circuit shown in Fig. P.7. *Hint*: Introduce a very small current source in parallel with the 60-Ω resistor to simulate no independent source in the original circuit.

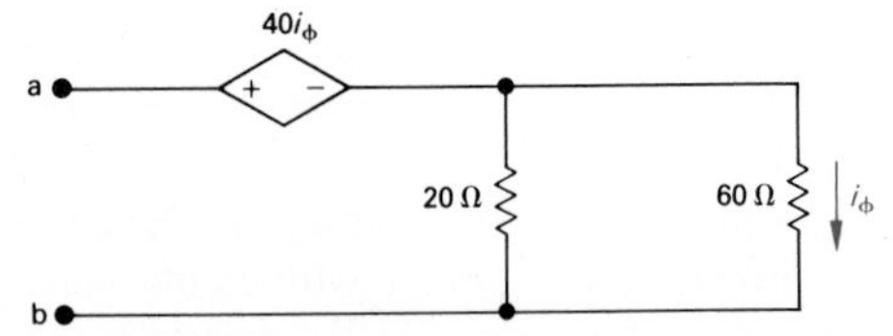

FIGURE P.7

8. Assume that the resistors in the ciruit shown in Fig. P.8 can vary by ±2% and that the voltage of the ideal voltage source can vary by ±5%. Use PSpice to predict the minimum and maximum values of v_o.

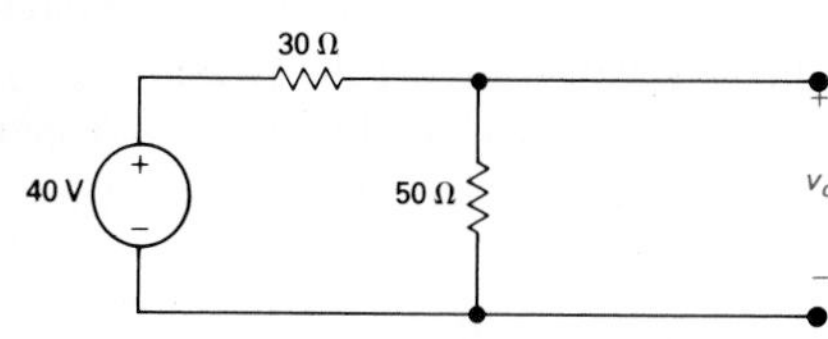

FIGURE P.8

9. The op amp in the noninverting amplifier circuit shown in Fig. P.9 has an input resistance of 40 kΩ, an output resistance of 1 kΩ, and an open-loop gain of 10^4.

a) Use PSpice to find the value of v_o and i_g if v_g = 1 V(dc).

b) Compare the PSpice solution with the analytic solution for v_o and i_g.

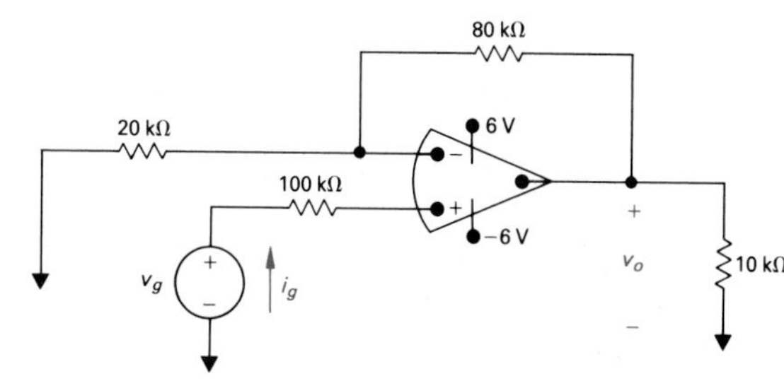

FIGURE P.9

10. The two op amps in the circuit shown in Fig. P.10 are identical. Their parameter values are R_i = 500 kΩ, R_o = 100 Ω, and $A = 10^4$.

a) Use PSpice to find v_{o1} and v_{o2} when v_g = 0.6 V(dc).

b) Calculate v_{o1} and v_{o2} if the amps are assumed ideal. Are these values of v_{o1} and v_{o2} consistent with the output from the PSpice simulation in (a)?

c) Now use the model of the μA741C operational amplifier found in the evaluation library to represent the op amps in Fig. P.10 and rerun the PSpice simulation. How do these results compare with those obtained in (a) and (b)?

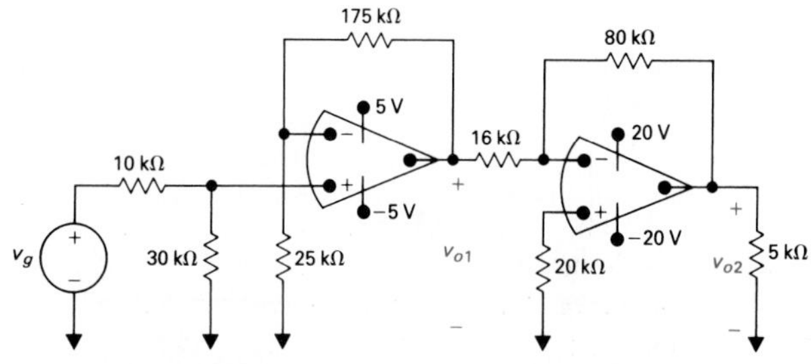

FIGURE P.10

11. An exponential voltage pulse is described by

$$v_g = 12.5 \text{ V}, \quad 0 \leq t \leq 1 \text{ s};$$
$$v_g = 12.5 + 12.5(1 - e^{-0.1(t-1)}) \text{ V}, \quad 1 \text{ s} \leq t \leq 11 \text{ s};$$
$$v_g = 12.5 + 12.5(1 - e^{-0.1(t-1)}) - 12.5(1 - e^{-0.125(t-11)}) \text{ V}, \quad 11 \text{ s} \leq t \leq 31 \text{ s}.$$

(continued)

These voltages are applied to the circuit shown in Fig. P.11.

a) Use PSpice and PROBE to obtain a plot of $v_o(t)$ for $0 \leq t \leq 31$ s at increments of 200 ms.

b) Use the output from PSpice analysis to check the validity of the analytic solution.

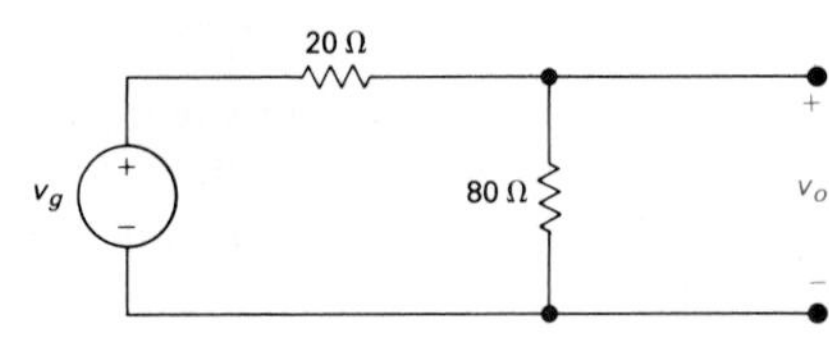

FIGURE P.11

12. An exponential voltage source is described by

$v_g = 10$ V, $\quad 0 \leq t \leq 1$ s;

$v_g = 10 + 30(1 - e^{-0.125(t-1)})$ V, $\quad 1$ s $\leq t \leq 3$ s;

$v_g = 10 + 30(1 - e^{-0.125(t-1)}) - 30(1 - e^{-(t-3)})$ V, $\quad 3$ s $\leq t \leq 20$ s.

a) Apply this pulse to the circuit shown in Fig. P.11 and use PSpice and PROBE to obtain a plot of $v_o(t)$ for $0 \leq t \leq 20$s in increments of 0.1 s.

b) Use the data from the PSpice simulation to verify the validity of the analytic solution.

13. The exponential pulse from Problem 12 is applied to the circuit shown in Fig. P.13.

a) Use PSpice and PROBE to obtain plots i_o and v_o for $0 \leq t \leq 10$ s.

b) Do the plots obtained in (a) make sense in terms of known circuit behavior? Explain.

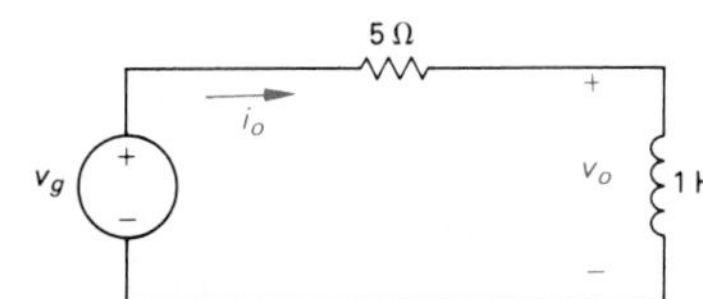

FIGURE P.13

14. Use PSpice and PROBE to plot i versus t for the circuit shown in Fig. P.14 when $I_o = 50$mA, $V_o = -10$ V, $L = 10$ mH, $R = 200\ \Omega$, and $C = 1\ \mu$F. Use an analytic solution for $i(t)$ to check the results displayed in the PROBE plot.

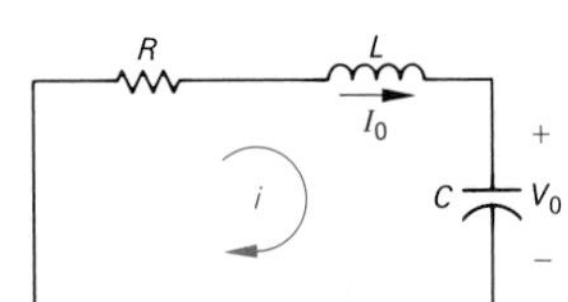

FIGURE P.14

15. a) In the circuit shown in Fig. P.15, $V_o = 0$ and $I_o = -12.25$ mA. The circuit element values are $R = 20$ kΩ, $L = 8$ H, and $C = 0.125\ \mu$F. Use PSpice and PROBE to plot $v(t)$ versus t for the time interval $0 \leq t \leq 11$ ms.

b) Use PROBE to identify the maximum value of $v(t)$, the time at which the maximum value occurs, and the value of the damped period.

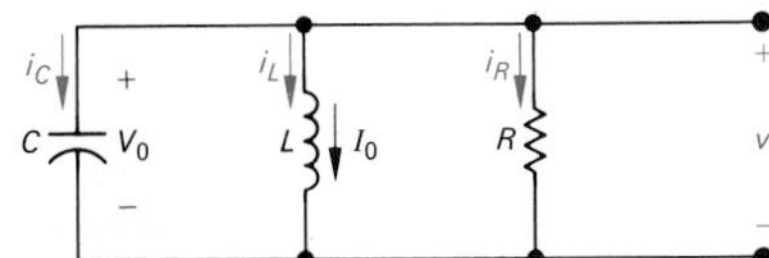

FIGURE P.15

16. a) Use PSpice and PROBE to plot $v(t)$ versus t for the circuit shown in Fig. P.15 when the 20 kΩ resistor is removed from the circuit.

b) Discuss the significance of removing the resistor from the circuit.

17. a) For the circuit shown in Fig. P.15, find the value of R that will result in a critically damped voltage response.

b) Use PSpice and PROBE to plot $v(t)$ versus t for $0 \leq t \leq 7$ ms.

c) Use PROBE to check the maximum value of $v(t)$ and the time at which it occurs.

18. The switch in the circuit shown in Fig. P.18 has been closed for a long time. At $t = 0$ it is opened. Use PSpice and PROBE to plot $i_L(t)$ versus t for $t > 0$. In the transient analysis, let tstop equal five time constants and select tstep to give 100 points between 0 and tstop.

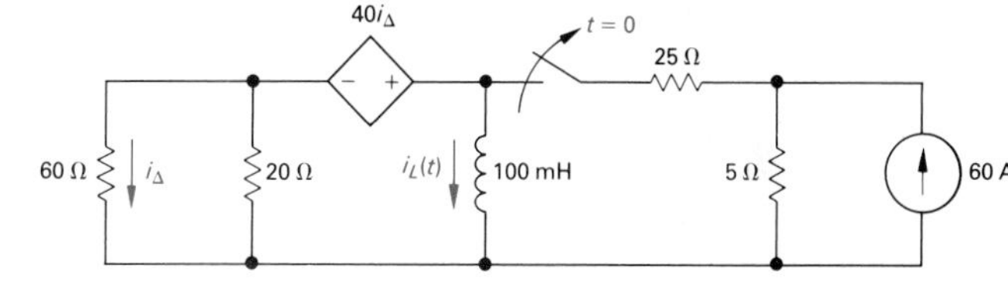

FIGURE P.18

19. The 360-V, 3.6-Ω source in the circuit shown in Fig. P.19 is inadvertently short-circuited at its terminals a, b. At the time the fault occurs, the circuit is in a steady-state operating condition.

a) Use PSpice and PROBE to obtain a plot of the short-circuit current for $t \geq 0$. Make a preliminary analysis of the circuit to facilitate the choice of tstop and tstep.

b) From the PSpice simulation and the PROBE plot, estimate at what time the short-circuit current equals 76 A.

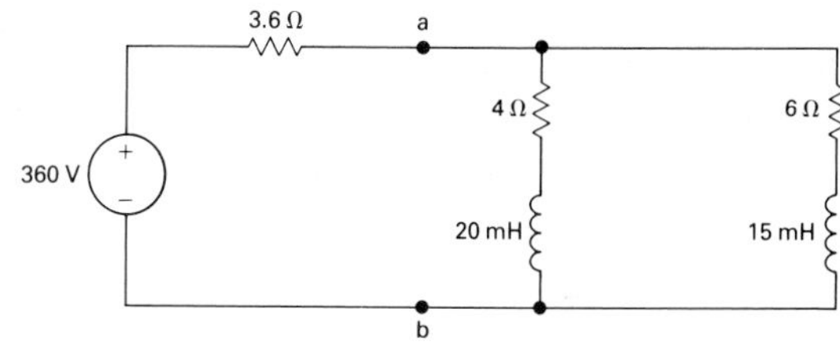

FIGURE P.19

20. The initial voltage on the capacitor in the circuit shown in Fig. P.20 is 100 V. Use PSpice simulation and PROBE to obtain a plot of v_C and v_ϕ for $t > 0$.

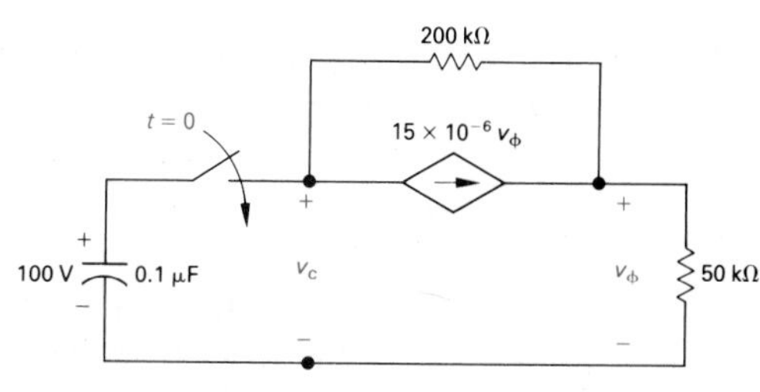

FIGURE P.20

21. a) The 0.1 μF capacitor in the circuit shown in Fig. P.21 is charged to 100 V. At $t = 0$ the capacitor is discharged through a series combination of a 100-mH inductor and a 560-Ω resistor. Use PSpice and PROBE to plot $i(t)$ and $v_C(t)$.

b) Use PROBE to confirm the analytic solutions for this circuit. In particular, check the peak values, time of occurrence, and damped period.

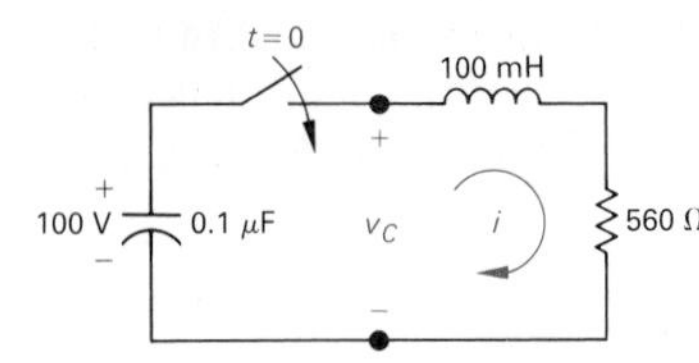

FIGURE P.21

22. a) No energy is stored in the 100-mH inductor or the 0.4-μF capacitor when the switch in the circuit shown in Fig. P.22 is closed. Use PSpice and PROBE to plot $v_C(t)$ versus t.

b) Use PROBE to confirm the analytic solution for $v_C(t)$.

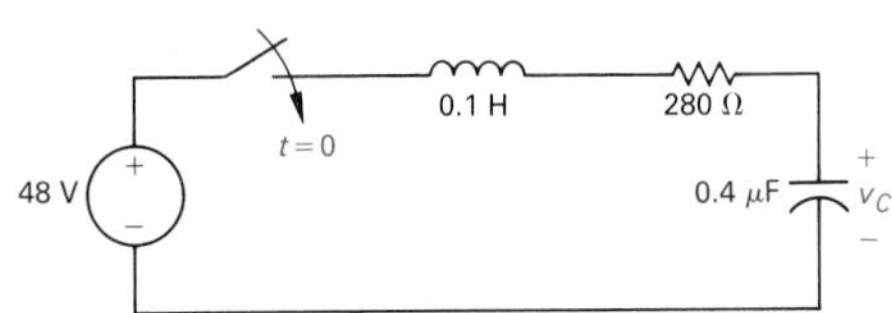

FIGURE P.22

23. The switch in the circuit shown in Fig. P.23 has been closed for a long time before opening at $t = 0$.

a) Calculate the damped period of the voltage $v_o(t)$.

b) Use PSpice and PROBE to obtain a plot of $v_o(t)$ over the interval $0 \leq t \leq 1.5T_d$.

c) Use PROBE to determine the maximum magnitude of $v_o(t)$ and the time at which it occurs.

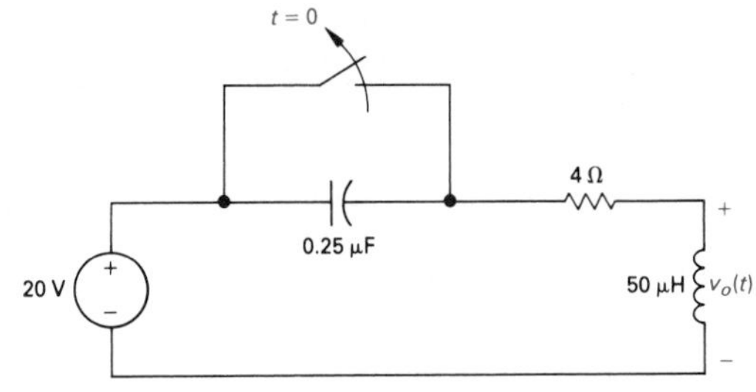

FIGURE P.23

24. No energy is stored in the circuit shown in Fig. P.24 at the instant the dc current source jumps from 0 to 8 A.

a) Use PSpice and PROBE to plot $i_L(t)$ versus t for $t \geq 0$.

b) Use PROBE to determine the peak value of $i_L(t)$ and the instant at which it occurs.

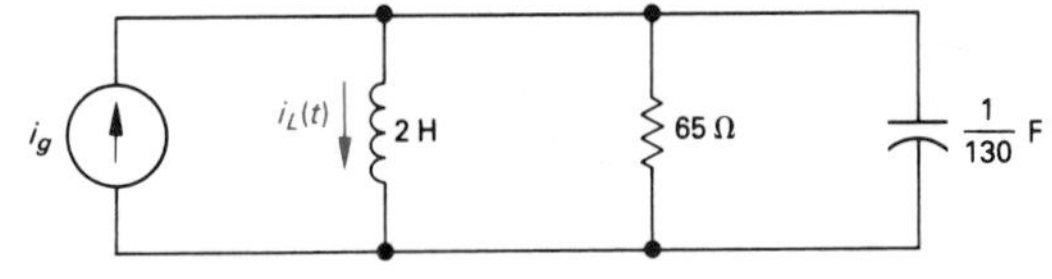

FIGURE P.24

25. a) No energy is stored in the circuit shown in Fig. P.25 at the time the sources are energized. Use PSpice and PROBE to plot v_2 versus t.

b) Use PROBE to find $v_2(0)$, $v_2(\infty)$, and $dv_2(0)/dt$.

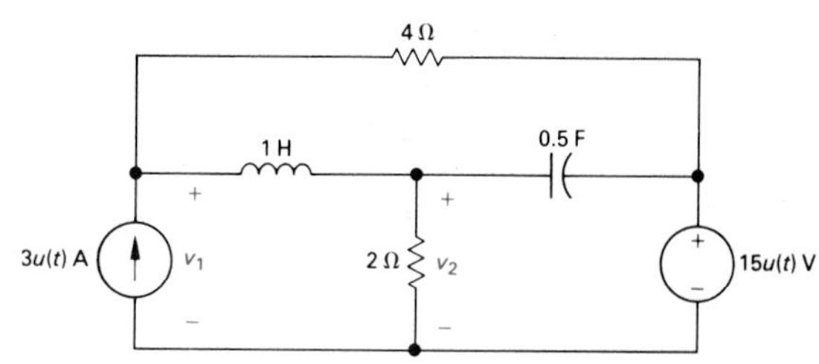

FIGURE P.25

26. a) The initial energy in the circuit shown in Fig. P.26 is zero. Use PSpice and PROBE to plot $i(t)$.

b) Use PROBE to determine $i(0)$, $i(\infty)$, and the damped period.

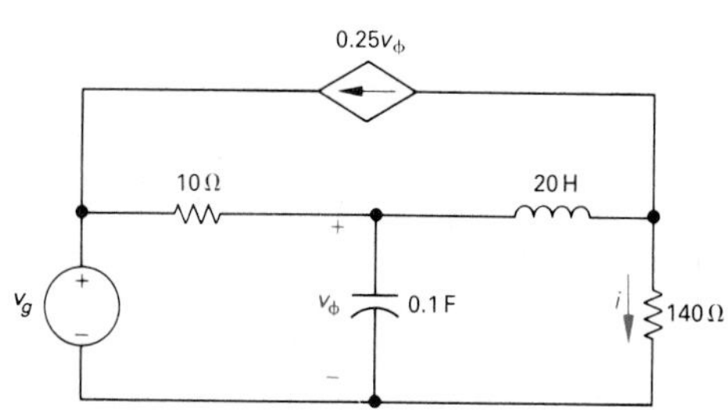

FIGURE P.26

27. a) No energy is stored in the circuit shown in Fig. P.27 when the current source jumps to 100 A. Use PSpice and PROBE to plot $i(t)$.

b) Use PROBE to determine $i(0)$ and $i(\infty)$.

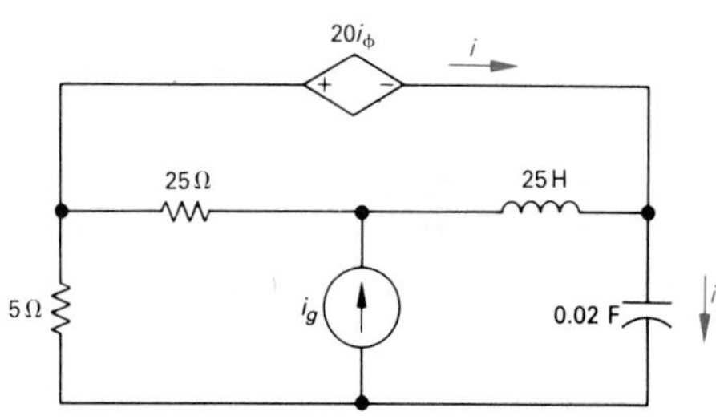

FIGURE P.27

28. a) Use PSpice to simulate the source file listed in Example 10 for the circuit shown in Fig. P.28.

b) Derive the analytic expression for $v_o(t)$ in the circuit of Example 10.

c) Compare the output from the PSpice simulation to results predicted by the analytic solution. For example, compare the results at $t = 0$, 40, and 80 ms and at ∞.

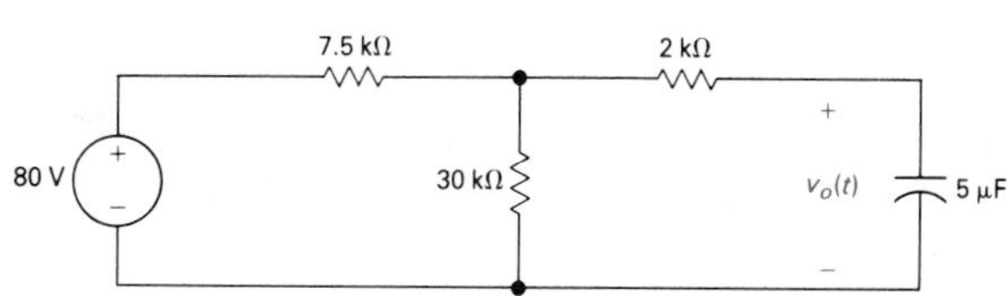

FIGURE P.28

29. The following parameters pertain to the circuit shown in Fig. P.29: $R = 1$ kΩ, $R_1 = 400$ kΩ, $R_2 = 100$ kΩ, $R_3 = 20$ kΩ, $R_4 = 5$ kΩ, $R_5 = 50$ kΩ, $R_6 = 0.5$ kΩ, $R_7 = 400$Ω, $R_8 = 50$ kΩ, $C_1 = 5\ \mu$F, and $C_2 = 5\ \mu$F. Assume that $f(t)$ is a dc voltage that switches from 0 to 4.04 V at $t = 0$. At the time the source is switched on, no energy is stored in the capacitors C_1 and C_2. The operational amplifiers are assumed to be ideal.

a) Use PSpice and PROBE to generate a plot of $v_B(t)$ and $v_C(t)$ versus t. Use the analytic solution to guide the choice of tstep and tstop.

b) Use PROBE to find the maximum value of $v_C(t)$ and the time at which it occurs.

c) Compare the results of (b) to the analytic solution for $v_C(max)$ and $t(max)$.

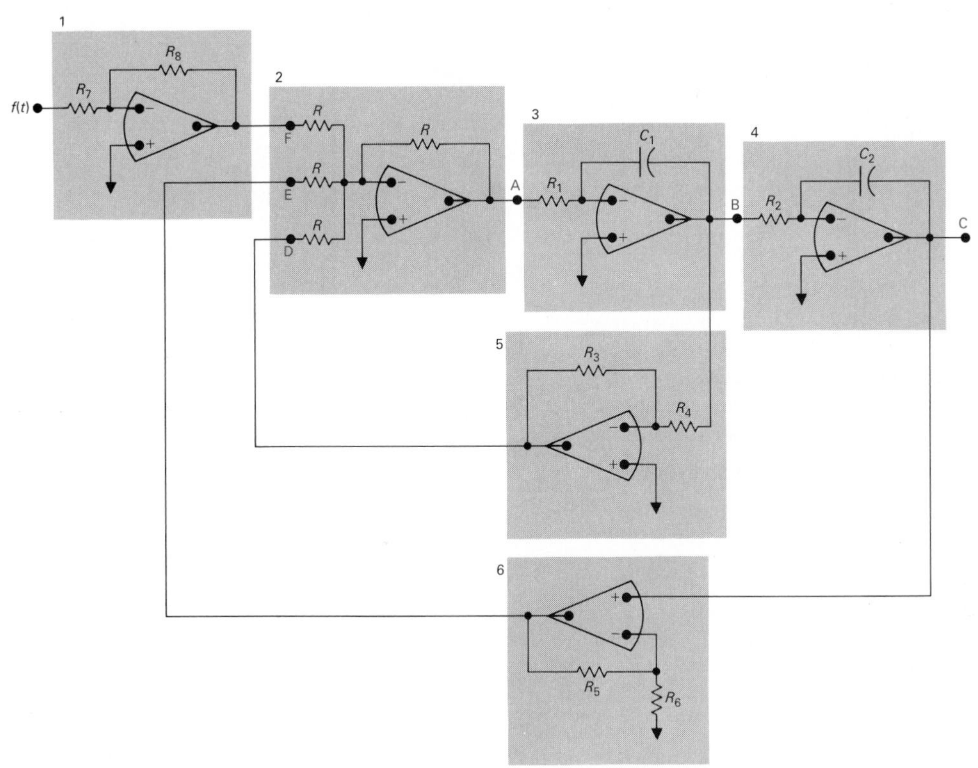

FIGURE P.29

30. a) Set the initial voltages on the capacitors C_1 and C_2 in the circuit of Problem 29 so that $v_C(0) = -5$ V and $dv_C(0)/dt = 10$ V/s.

b) Use PSpice and PROBE to obtain a plot of v_C and v_B with the initial conditions from (a).

c) Use PROBE to check on whether $v_C(0)$, $dv_C(0)/dt$, and $v_C(\infty)$ are satisfied.

d) Use the output from PSpice and PROBE to estimate T_d and check it against the calculated value.

31. Replace two of the ideal op amps in Problem 29 with the μA741C op-amp model from the nominal library. Use PSpice and PROBE to plot $v_C(t)$ versus t and compare your results to those in Problems 29 and 30.

32. The op amps in the circuit shown in Fig. P.32 are ideal. At the instant the voltage source jumps from 0 to 250 mV no energy is stored in the capacitors.

a) Use PSpice and PROBE to plot v_{o1} and v_{o2} for $0 \leq t \leq 0.5$ s.

b) Compare the results to the analytic solution for v_{o1} and v_{o2} at $t = 0, 0.10, 0.20, 0.30, 0.40$, and 0.50 s.

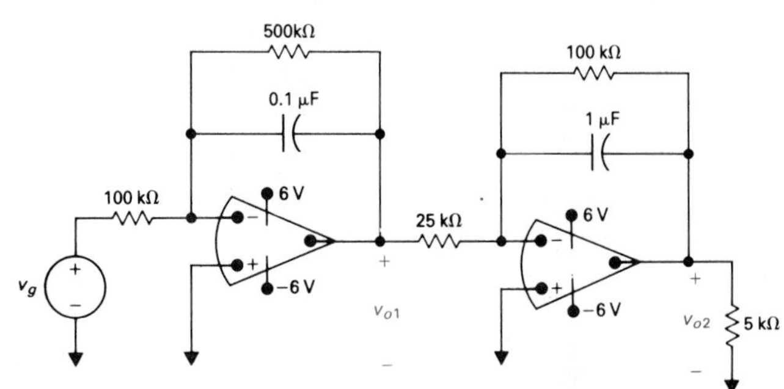

FIGURE P.32

33. Use PSpice together with PROBE to plot the transient response of v_C in the circuit shown in Fig. P.33 from 300 to 1800 μs in steps of 25 μs. Use a 5-μs increment for calculating the transient response.

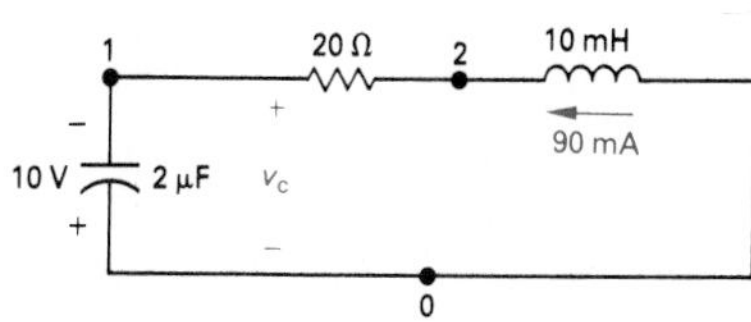

FIGURE P.33

34. Use the . STEP statement in a PSpice source file to increase the size of the capacitor in the circuit shown in Fig. P.23 from 0.25 μF to 12.5 μF in steps of 0.5 μF. Use PROBE to find the maximum amplitude of $v_o(t)$ and the time at which it occurs as the capacitor value changes.

35. a) For the damped sinusoidal voltage

$$v_g = 12.5 + 25e^{-0.1(t-1)}\sin\,[0.4\pi\,(t - 1)]\text{ V}$$

calculate the maximum value of v_g, time at which the maximum occurs, damped period, delay time, and offset voltage.

b) Use this sinusoid as the input to the circuit shown in Fig. P.35 and use PSpice to simulate this circuit.

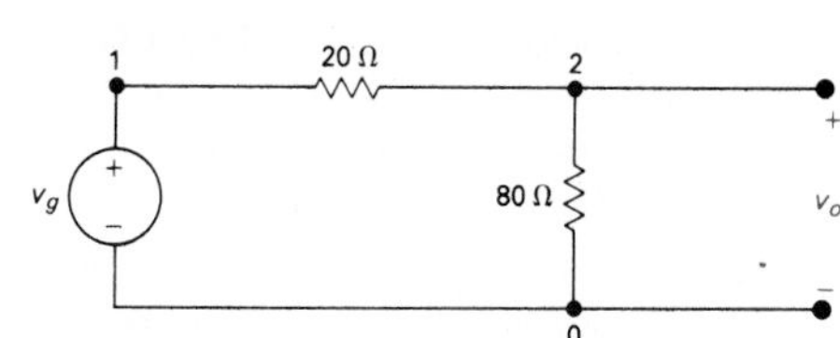

FIGURE P.35

(continued)

c) From the output of PSpice simulation, tabulate the maximum value of v_g, time at which the maximum occurs, damped period, delay time, and offset voltage.

d) Compare the calculated values from (a) to the output from PSpice simulation in (c).

36. Assume that the damped sinusoidal source description in Problem 35 is applied to the circuit shown in Fig. P.36. At the time v_g is energized, the initial charge on the capacitor is zero.

a) Use PSpice and PROBE to plot $v_o(t)$ versus t.

b) Does the plot obtained in (a) make sense in terms of known circuit behavior?

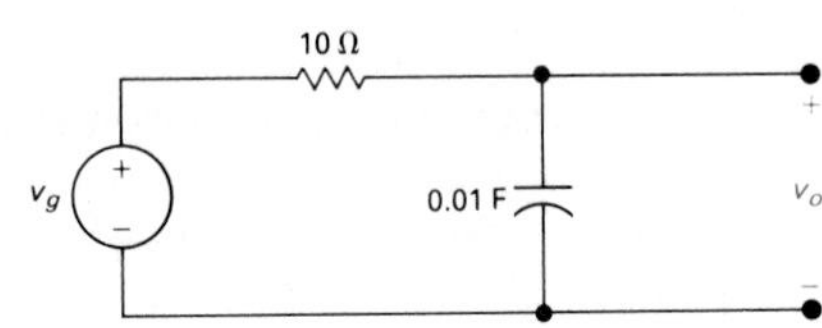

FIGURE P.36

37. The sinusoidal voltage source in the circuit shown in Fig. P.37 is generating a voltage of $6 \cos 10^4 \pi t$ V.

a) Use PSpice to find the amplitude and phase angle of v_0, v_1, v_2, and i_0.

b) Test the results obtained in (a) by showing that they are consistent with the conservation of energy principle—namely, that the average power delivered equals the average power absorbed.

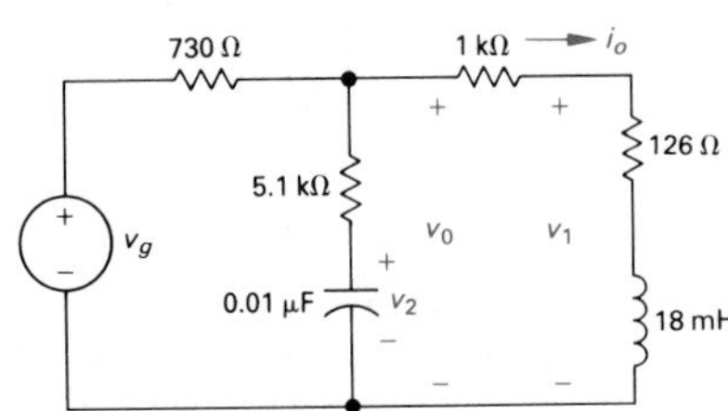

FIGURE P.37

38. The phasor-domain circuit shown in Fig. P.38 is based on a frequency of 60 Hz.

a) Use PSpice to find the amplitude and phase angle of V_1, V_2, and V_3.

b) Repeat (a) with the impedance of the neutral conductor first increased by a factor of 100 (that is, $Z_n = 6 + j6\ \Omega$) and then increased by a factor of 1000 (that is, $Z_n = 60 + j60\ \Omega$).

c) Describe the adverse effect on the operation of the circuit when the impedance of the neutral conductor is increased.

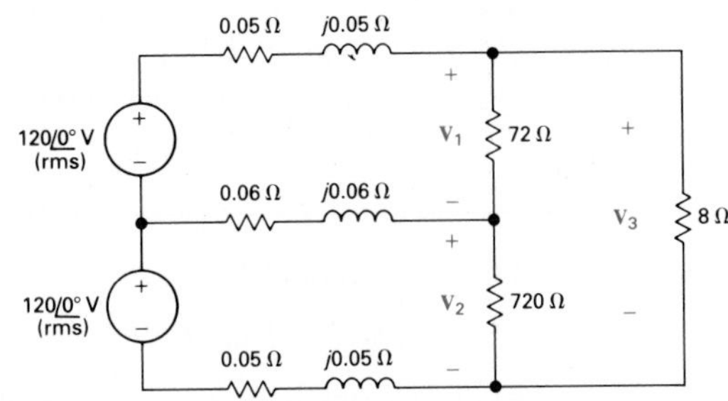

FIGURE P.38

39. The phasor-domain circuit shown in Fig. P.39 is based on a frequency of 5000 rad/s. Use PSpice to find the magnitude and phase angle of I_o and I_1 if $V_g = 40 + j30$ V.

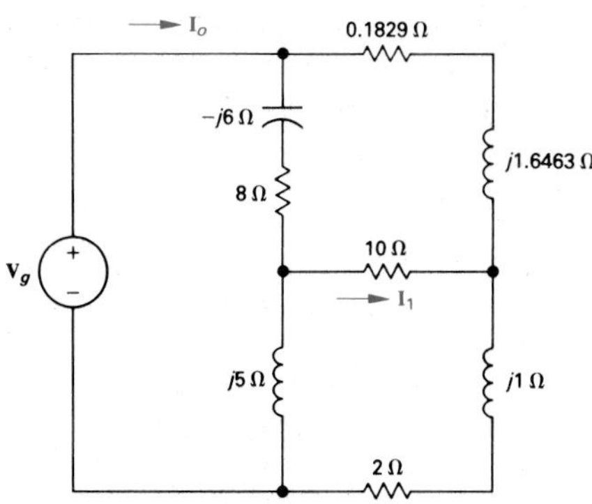

FIGURE P.39

40. An electrical load operates at 240 V(rms) with a frequency of 60 Hz. The load absorbs an average power of 8 kW at a lagging power factor of 0.8.

a) Calculate the reactive power of the load and the complex power of the load.

b) Calculate the impedance of the load and use this impedance to construct a PSpice source file.

c) Simulate the circuit using the source file from (b). Then use PROBE to generate plots of the instantaneous power, average power, reactive power, and complex power.

d) How do the PROBE results compare to the analysis in (a)?

41. a) For the circuit shown in Fig. P.41, use PSpice to find the value of capacitance that results in transfer of the maximum average power to the load. Assume that the load resistance is 4000 Ω and that the source frequency is 60 Hz.

b) For the capacitor value determined in (a), use PROBE to compute the average power transferred to the load. Does this agree with the analytic result?

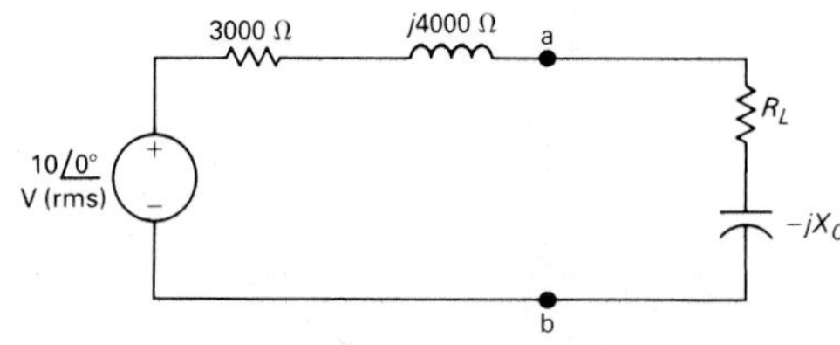

FIGURE P.41

42. For the circuit shown in Fig. P.41, assume that the load resistance can vary from 0 to 4000 Ω and that the load capacitive reactance can vary from 0 to −2000 Ω. Use PSpice and PROBE to determine the value of load resistance and load capacitance that transfers the most average power to the load. Assume that the source operates at a frequency of 60 Hz.

43. The sinusoidal voltage source in the circuit shown in Fig. P.43 is described by the equation $v_g = 202 \cos 75000t$ V. Use PSpice to find the amplitude and phase angle of i_o.

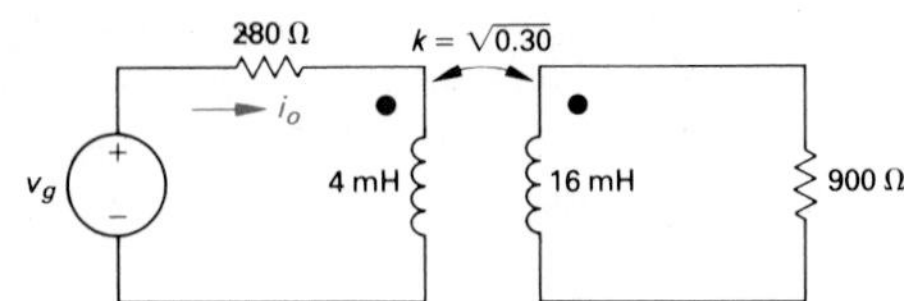

FIGURE P.43

44. Use PSpice to find the amplitude and phase angle of i_g in the circuit shown in Fig. P.44 when $v_g = 100 \cos 2t$ V.

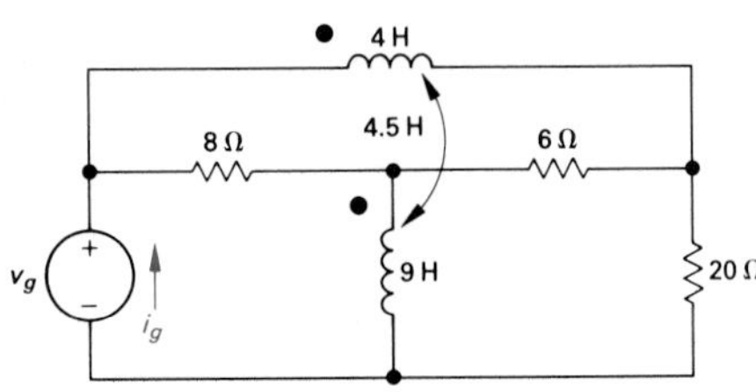

FIGURE P.44

45. a) Use PSpice to find the rms magnitude and the phase angle of i_1 and i_2 in the circuit shown in Fig. P.45. The sinusoidal voltage source is operating at a frequency of 1600 rad/s and has an rms value of 150 V.

b) Test the PSpice simulation results by using them to determine whether average power generated equals average power dissipated.

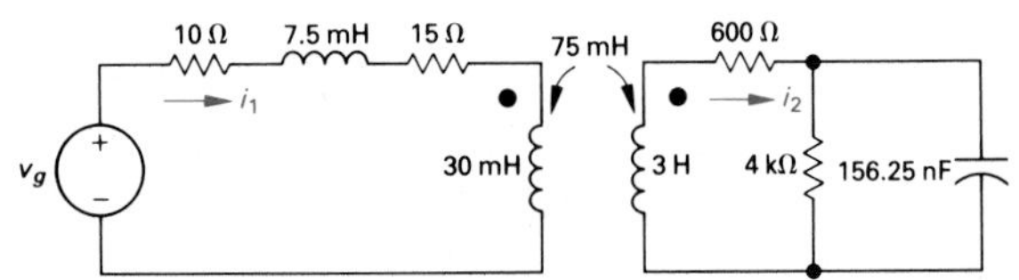

FIGURE P.45

46. The coefficient of coupling k in the circuit shown in Fig. P.46 is varied from 0.3 to 1.0 in steps of 0.1.

a) Use PSpice to construct a table that shows how the magnitude of i_1 varies with k when $v_g = 100 \cos 75000t$ V.

b) Analytically solve for the value of k that yields the maximum magnitude of i_1. Call this value k_{max}.

c) Use PROBE to predict the value of k_{max} from the simulation and compare that with the value from the analysis.

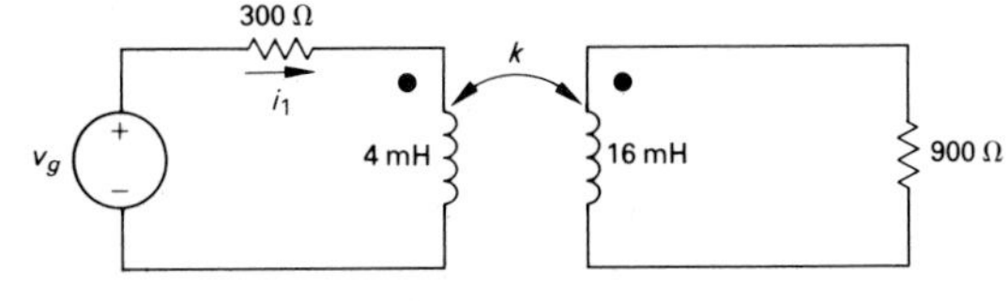

FIGURE P.46

47. a) Use PSpice to find the magnitude and phase angle of v_o in the circuit shown in Fig. P.47 if $v_g = 500 \cos 1000\pi t$ V.

b) Compare the results of PSpice simulation to the analytic solution for v_o.

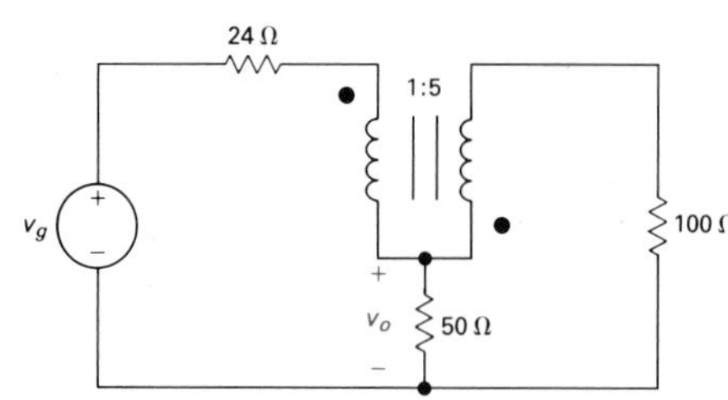

FIGURE P.47

48. a) Use PSpice to find the magnitude and phase angle of v_o in the circuit shown in Fig. P.48 if $v_g = 456 \cos 5000t$ V, $N_1 = 1000$ turns, and $N_2 = 500$ turns.

b) Compare the PSpice simulation results to the analytic solution for v_o.

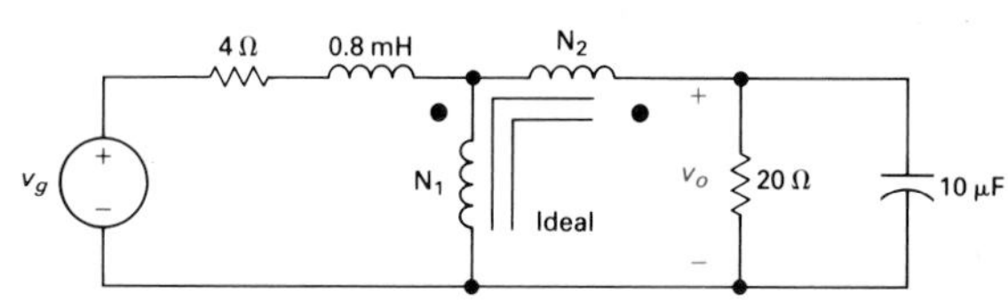

FIGURE P.48

49. a) Use PSpice and PROBE to plot the amplitude of v_o versus frequency for the circuit shown in Fig. P.49.

b) Use PROBE to estimate the resonant frequency, bandwidth, and quality factor of the circuit.

c) Compare the results obtained in (b) to calculated values of f_o, β, and Q.

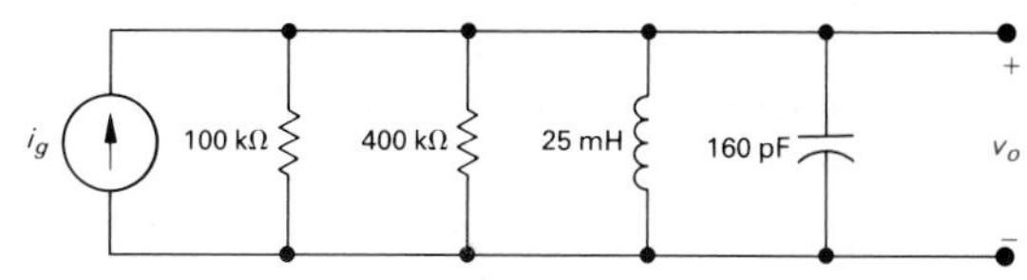

FIGURE P.49

50. a) The current source in the parallel resonant circuit shown in Fig. P.50 is $i_g =$ 4 cos ωt mA. Calculate the unity power factor resonant frequency (f_o), the frequency at which v_o is maximum (f_m), the value of v_o at f_o, and the value of v_o at f_m.

b) Use PROBE to plot the amplitude of v_o versus f.

c) Compare the results of PSpice simulation to the results from (a).

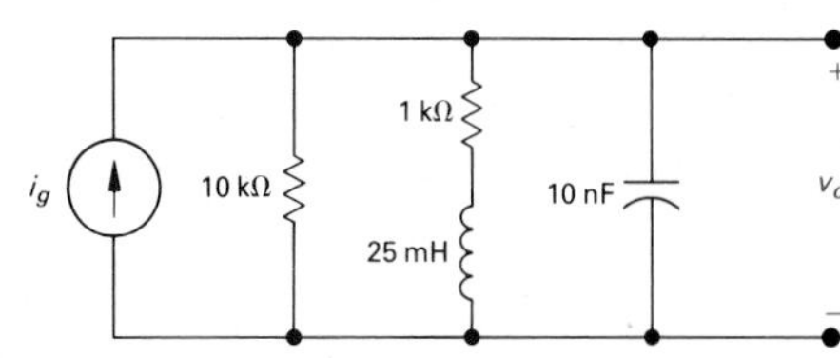

FIGURE P.50

51. The sinusoidal voltage source (v_g) in the circuit shown in Fig. P.51 has a peak amplitude of 10 V. Use PSpice and PROBE to identify (a) the unity power factor resonant frequency; (b) the magnitude of v_o at unity power factor resonance; (c) the maximum magnitude of v_o; and (d) the frequency at which the maximum magnitude occurs.

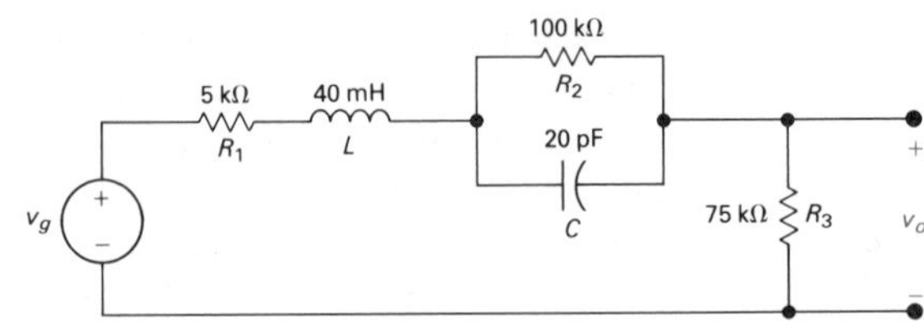

FIGURE P.51

52. The circuit parameters shown in Fig. P.51 are changed to the following values: $R_1 = 2$ kΩ, $L = 20$ mH, $R_2 = 200$ kΩ, $C = 250/21$ pF, and $R_3 = 8$ kΩ. At the same time the peak amplitude of v_g is changed to 100 mV.

a) Derive the transfer function V_o/V_g.

b) Calculate the unity power factor resonant frequency f_o.

c) Calculate the magnitude of v_o in decibels at the frequency found in (b).

d) Calculate the frequency (f_m) at which the magnitude of v_o is maximum.

e) Calculate the magnitude of v_o in decibels at the frequency found in (d).

f) Use PSpice and PROBE to construct a plot of v_o in decibels versus $\log f$. Start the plot at $f_m/1000$ and stop the plot at $10f_m$. Overlay a straight-line Bode magnitude plot on a hard copy of the PROBE plot and compare.

g) Use the PROBE cursor to identify the values calculated in (b)–(d). How do the simulation results compare to the analytic results?

53. In the circuit shown in Fig. P.53 the sinusoidal voltage source is generating the voltage $v_g = 100 \cos 64500t$ V. The variable capacitor is varied from 5.72 nF to 9.72 nF in steps of 1 nF. Use the . STEP statement in PSpice to simulate this circuit and PROBE to plot the steady-state magnitude of v_o versus C for the stepped values of C.

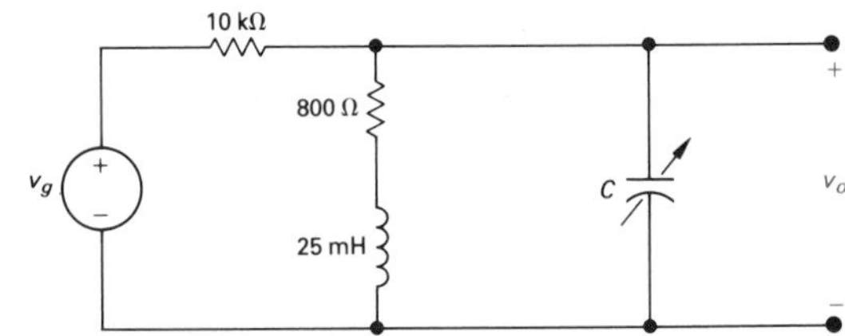

FIGURE P.53

54. For the circuit shown in Fig. P.53, use PSpice to simulate the frequency response characteristics of the circuit. Continue to use the . STEP statement to vary the capacitance but now let the frequency vary from 100 Hz to 10^6 Hz. Use PROBE to graph the results in Bode plot form: That is, plot the magnitude of v_o in decibels versus $\log f$. Overlay a straight-line Bode magnitude plot using the PROBE label menu and comment on its accuracy.

55. a) For the circuit shown in Fig. P.55, derive the transfer function V_o/I_g.

b) Use PSpice and PROBE to produce a Bode-type frequency plot, including both the magnitude of v_o in decibels and the phase angle of v_o in degrees versus log f. Be sure to choose a range of decades for the PSpice simulation that includes the frequencies where the magnitude and phase angle change.

c) Overlay the straight-line Bode approximations on both the magnitude and the phase angle plots produced by PROBE. How close are the Bode plots to the actual plots? For what range of frequencies are the two plots most different?

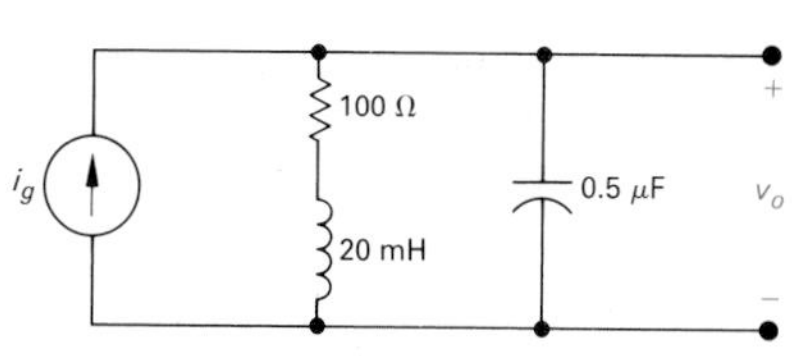

FIGURE P.55

56. The circuit shown in Fig. P.56 has a voltage source of constant magnitude and phase angle but variable frequency. Use PSpice and PROBE to create a family of Bode-type frequency response plots of the magnitude of v_o in decibels and the phase angle of v_o in degrees versus log f as R_x is varied over the range 0 Ω to 10^6 Ω. Choose about six values for R_x within that range. Then overlay the Bode straight-line magnitude and phase angle plots on those produced by PROBE. For what value of R_x does the Bode plot vary the least from the actual plot? For what value of R_x does the Bode plot vary the most from the actual plot?

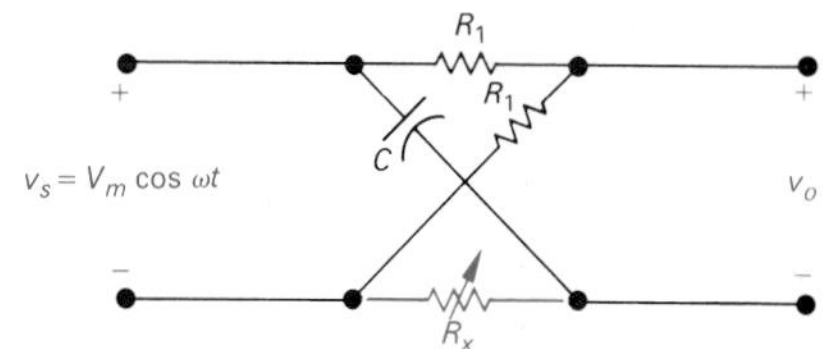

FIGURE P.56

57. a) Derive the analytic solution for the circuit shown in Fig. P.57, where $i_g = 7\,\delta(t)$.

b) From the analytic solution, find the minimum value of $v_o(t)$, the time at which the minimum occurs, and the damped period.

c) Use PSpice to simulate this same circuit.

d) Use PROBE to plot $v_o(t)$ versus t and determine from the plot the values from (b). How do the simulated values compare to the analytic values?

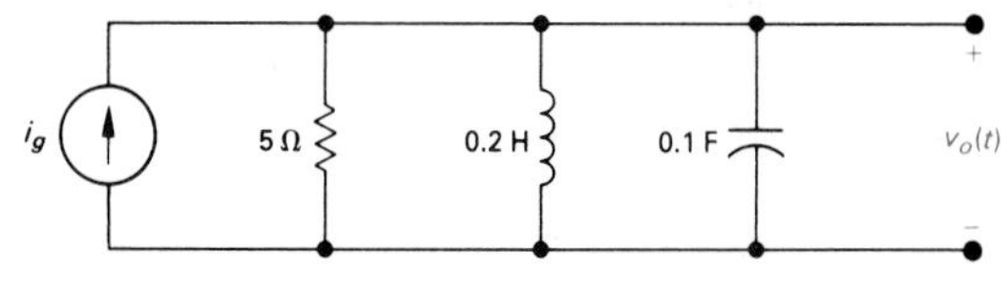

FIGURE P.57

58. The voltage source v_g in the circuit shown in Fig. P.58(a) is generating the signal shown in Fig. P.58(b).

a) Derive the expressions for $v_o(t)$ for the time intervals $t < 0$, $0 \leq t \leq 0.5$ s, $0.5\ \text{s} \leq t \leq 1.0$ s, and $t \geq 1.0$ s.

b) Use PSpice and PROBE to obtain a plot of $v_o(t)$ versus t. Simulate the voltage source with a pulse.

c) Compare the results from the simulation with those predicted by the analytic expressions.

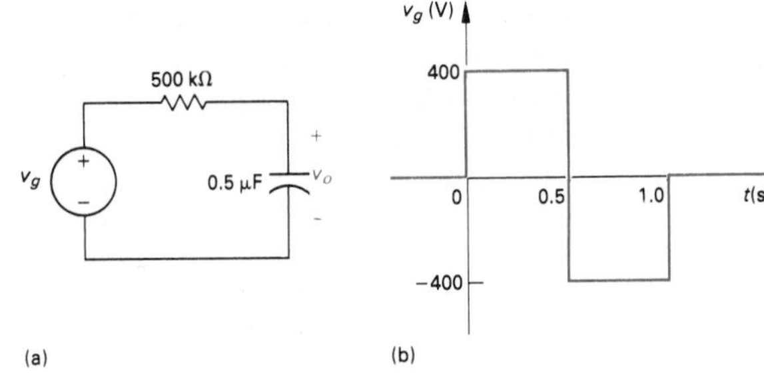

FIGURE P.58

59. The current source i_g in the circuit shown in Fig. P.59(a) has the waveform shown in Fig. P.59(b).

a) Use PSpice and PROBE to plot $i_o(t)$ versus t for $0 \leq t \leq 1000\ \mu$s.

b) Derive the analytic expressions for $i_o(t)$ for $t \geq 0$.

c) Use the results from PSpice simulation to check the results predicted by the analysis in (b).

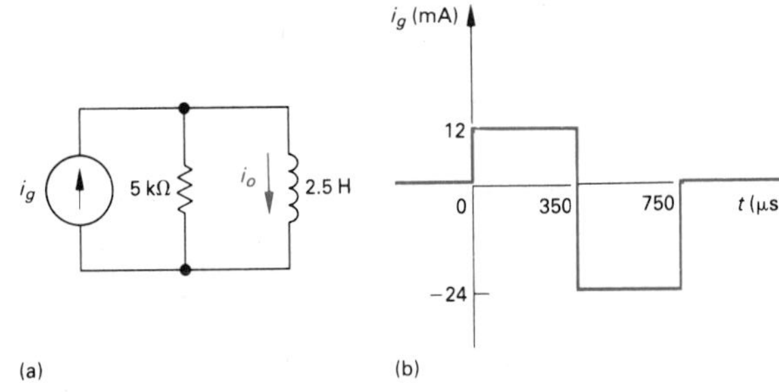

FIGURE P.59

60. The circuit shown in Fig. P.60 has been in operation a long time. At $t = 0$, the voltage drops instantaneously to 12.5 V and the current source reverses direction.

a) Use PSpice and PROBE to obtain a plot of $v_o(t)$ versus t for $0 \leq t \leq 500$ ms.

b) From the PROBE plot, note the time at which $v_o(t)$ is zero.

c) Check the result in (b) with the analytic solution for $v_o(t) = 0$.

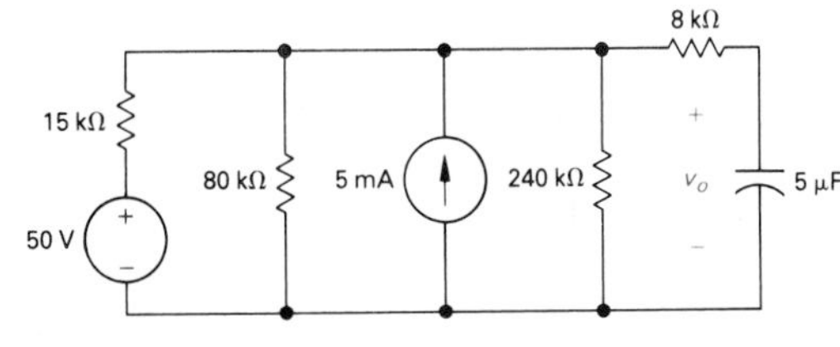

FIGURE P.60

61. There is no charge on the capacitor when the voltage pulse shown in Fig. P.61(a) is applied to the ideal integrating amplifier circuit shown in Fig. P.60.

a) Use PSpice and PROBE to obtain a plot of $v_o(t)$ versus t over the time interval 0 to 3 s.

b) Using the PROBE cursor, read off the values of v_o at 0, 0.5, 1.0, 1.5, 2.0, 2.5, and 3 s.

c) Compare the simulation results to the analytic solution for v_o.

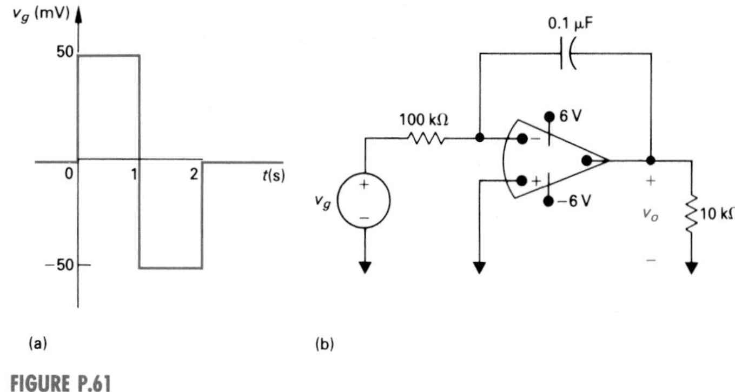

FIGURE P.61

62. Repeat Problem 61 with the feedback capacitor paralleled by a 5-MΩ resistor. What is the effect of the parallel resistor?

63. Assume that the ideal operational amplifier in the circuit shown in Fig. P.61(b) is replaced by an op amp that has an output resistance of 75 Ω, an input resistance of 2 MΩ, and an open-loop gain of 10^5.

a) Use PSpice and PROBE to plot the output voltage in the interval $0 \leq t \leq 3$ s.

b) Compare the output to the ideal operational amplifier circuit simulation, considering especially the solution at 0, 0.5, 1.0, 1.5, 2.0, 2.5, and 3 s.

64. The signal voltage in the circuit shown in Fig. P.64(a) is the periodic square wave shown in Fig. P.64(b).

a) Assume that $v_{o1}(0) = 4$ V and use PSpice and PROBE to obtain a plot for the interval $0 \leq t \leq 1.6$ s.

b) Compare the results of the PSpice simulation of v_{o1} to the analytic solution for v_{o1}.

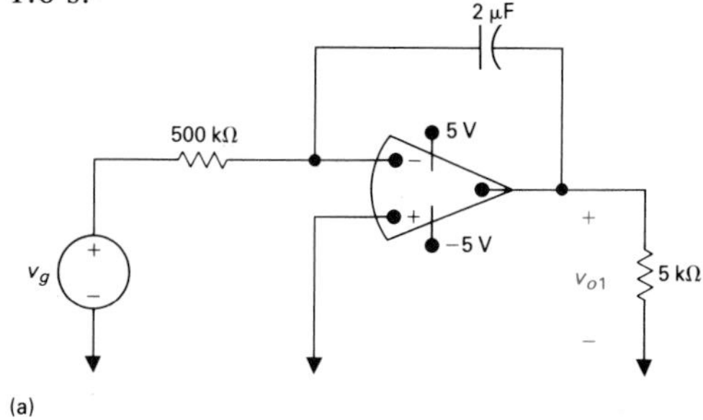

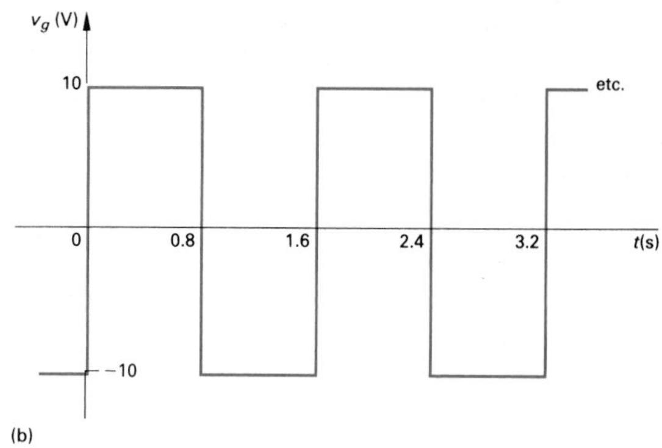

FIGURE P.64

65. The input signal in the circuit shown in Fig. P.65 is the periodic square wave shown in Fig. P.64(b).

a) Assume that $v_{o1}(0) = 4$ V and $v_{o2}(0) = 0$ V. Use PSpice and PROBE to obtain a plot of v_{o1} and v_{o2} for $0 \leq t \leq 1.8$ s.

b) Compare the PSpice simulation for v_{o2} to the analytic solution for v_{o2}.

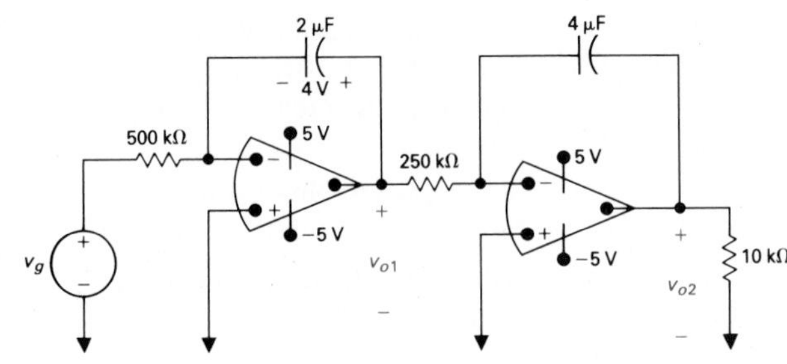

FIGURE P.65

66. a) Compute the expressions for the Fourier Series coefficients of the periodic waveform shown in Fig. P.64(b).

b) Compute the expressions for the Fourier Series coefficients of v_{o1} in the circuit shown in Fig. P.64(a). Assume that its input is the waveform shown in Fig. P.64(b).

c) Use the PSpice . FOUR statement to compute the Fourier Series coefficients of the first nine harmonic components for both the input waveform and the output voltage. How do the analytic results compare to the PSpice outputs?

d) Use PROBE to plot v_g and v_{o1} versus t. Then use PROBE to compute and display the Fourier transforms of these waveforms. How do they compare? What is the effect of the op-amp circuit on the frequency content of the voltage at the output?

67. a) Compute the expressions for the Fourier Series coefficients of v_{o1} and v_{o2} in the circuit shown in Fig. P.65. Assume that its input is the waveform shown in Fig. P.64(b).

b) Use the PSpice . FOUR statement to compute the Fourier Series coefficients of the first nine harmonic components for the input waveform, v_{o1}, and v_{o2}. How do the analytic results compare to the PSpice outputs?

c) Use PROBE to plot v_g, v_{o1}, and v_{o2} versus t. Then use PROBE to compute and display the Fourier transforms of these waveforms. How do they compare? What is the effect of the op-amp circuit on the frequency content of v_{o1} and v_{o2}?

ANSWERS TO SELECTED PROBLEMS

1. 198 V; 66 V; −132 V

4. 111.20 V; 117 V; 228.3 V

7. $V_t = 0$ V; $R_t = 25\Omega$

14. The PROBE plot agrees with the solution
$i = (500t + 0.05)e^{-10^4 t}$ A, $t \geq 0$.

19. a) The PROBE plot agrees with the solution
$i = 100 - 36e^{-200t} - 24e^{-400t}$ A.

b) 3.47 ms

23. a) 22.44 μs

b) The PROBE plot agrees with the analytic solution $(-500/7)e^{-\alpha t} \sin \omega_d t$, where $\alpha = 40000$ rad/s, $\omega_d = 280$ krad/s.

c) 67 V at 3.4 μs

24. a) $i_L = 8 - 8e^{-t} \cos 8t - e^{-t} \sin 8t$ A

b) From the plot, $i_L(max) = 13.39$A at 400 ms.

27. a) The PROBE plot agrees with the analytic solution $(-100 + 84e^{-t} + 36e^{-2t})u(t)$ A.

b) $i(0) = 20$ A; $i(\infty) = -100$ A

29. a) The PROBE plots agree with the analytic solution for v_B and v_C: $v_C = (5 - 5e^{-t}\cos 10t - 0.5e^{-t}\sin 10t)$ V and $v_B = -25.25e^{-t}\sin 10t$ V.

b) 8.655 V; 320 ms

c) 8.652 V; 314.16 ms; compare to (b)

32. a) The PROBE plots agree with the analytic solutions: $v_{o1} = (-1.25 + 1.25e^{-20t})$ V, $t \geq 0$ and $v_{o2} = (5 - 10e^{-10t} + 5e^{-20t})$ V, $t \geq 0$.

b)

t	v_{o1} (V)		v_{o2} (V)	
	PROBE	**Anal.**	**PROBE**	**Anal.**
0	0(*)	0	0(*)	0
0.1	−1.081	−1.081	1.996	1.998
0.2	−1.227	−1.227	3.738	3.738
0.3	−1.247	−1.247	4.515	4.515
0.4	−1.25	−1.2496	4.819	4.819
0.5	−1.25	−1.2499	4.933	4.933

(*) Actually, the cursor reading from PROBE for this value is not exactly zero, due to its representation in the computer, but has a value whose units are microvolts (μV), which is essentially zero.

35. a) 34.63 V; 2.19 s; 5 s; 1 s; 12.5 V

c) 34.59 V; 2.20 s; 5 s; 1 s; 12.5 V

d) Compare the values in (a) and (c).

37. a) $v_o = 3.675\angle 6.886°$ V; $v_1 = 1.690\angle 57.66°$ V; $v_2 = 1.946\angle -51.14°$ V; $i_o = 2.917\angle -19.78°$ mA

b) $P_{del} = \sum P_{dis} = 9.6637$ mW

39. $I_o = 18.68\angle -9.9°$ A; $I_1 = 3.688\angle 139.4°$ A

45. a) $i_1 = 2.361\angle - 64.29°$ A;
$i_2 = 74.15\angle - 21.41°$ mA

b) $P_{del} = 154.64$ W; $P_{dis} = 153.65$ W; checks to 1 W in 155 W

48. a) $508.4\angle - 41.99°$ V;

b) $508.41\angle - 41.99°$ V

51. a) 159.155 kHz;

b) 7.50 V;

c) 7.94 V;

d) 225.25 kHz

53.

C (nF)	**$\lvert v_o \rvert$ (V)**
5.72	27.03
6.72	28.34
7.72	28.83
8.72	28.34
9.72	27.01

58. a) $v_o = 0, t < 0$
$v_o = 400 - 400e^{-4t}$ V, $0 \leq t \leq 0.50$ s
$v_o = -400 + 745.87e^{-4(t-0.5)}$ V, $0.5 \leq t \leq 1$ s
$v_o = -2.99.06e^{-4(t-1)}$ V, $1 \leq t \leq \infty$ s

b) The PROBE plot agrees with the analytic expressions in (a).

c)

Quantity	**Anal. (V)**	**PROBE (V)**
$v_o(0.2)$	220.27	219.7
$v_o(0.5)$	345.87	345.0
$v_o(0.7)$	−64.86	−64.41
$v_o(1.0)$	−299.06	−298.60
$v_o(1.2)$	−134.38	−134.30

61. a) The PROBE plot agrees with the analytic solution.

b), c)

t (s)	PROBE v_o (V)	Anal. v_o (V)
0	4.2×10^{-8}	0
0.5	-2.497	-2.5
1.0	-4.994	-5.0
1.5	-2.497	-2.5
2.0	7.975×10^{-5}	0
2.5	7.975×10^{-5}	0
3.0	7.975×10^{-5}	0

65. a) The PROBE plots agree with the analytic solutions:

$v_{o1} = (-10t + 4)$ V, $0 \leq t \leq 0.8$ s
$v_{o1} = (10t - 12)$ V, $0.8 \leq t \leq 1.6$ s
$v_{o1} = (-10t + 20)$ V, $1.6 \leq t \leq 2.4$ s

$v_{o2} = (5t^2 - 4t)$ V, $0 \leq t \leq 0.8$ s
$v_{o2} = (-5t^2 + 12t - 6.4)$ V, $0.8 \leq t \leq 1.6$ s
$v_{o2} = (5t^2 - 20t + 19.2)$ V, $1.6 \leq t \leq 2.4$ s

b)

t	PROBE v_{o2} (V)	Anal. v_{o2} (V)
0	-2.5×10^{-8}	0
0.2	-0.60	-0.60
0.4	-0.80	-0.80
0.6	-0.60	-0.60
0.8	2.857×10^{-8}	0
1.0	0.60	0.60
1.2	0.80	0.80
1.4	0.60	0.60
1.6	-1.722×10^{-5}	0

INDEX